BACK IN ORBIT

JOHN GLENN'S RETURN TO SPACE

Published by LONGSTREET PRESS, INC.,
a subsidiary of Cox Newspapers,
a subsidiary of Cox Enterprises, Inc.
2140 Newmarket Parkway
Suite 122
Marietta, Georgia 30067

Printed in the United States of America

1st printing, 1998

Library of Congress Catalog Card Number: 98-066358

ISBN: 1-56352-525-9

Book and jacket design by Burtch Bennett Hunter

BACK IN ORBIT

JOHN GLENN'S RETURN TO SPACE

SCOTT MONTGOMERY
and
TIMOTHY R. GAFFNEY

of the
Pulitzer Prize-winning
Dayton Daily News

Foreword by Mercury astronaut Scott Carpenter

LONGSTREET
Atlanta, Georgia

Dedicated to two Dayton bicycle-makers,
Orville and Wilbur Wright, who taught the world to fly,
and to the courageous men and women since who have taken us higher and faster.

ACKNOWLEDGMENTS

In any major work, there are behind-the-scenes contributors who deserve recognition. Without their help, we couldn't have completed this nearly year-long project.

Dr. David R. Liskowsky, NASA program scientist for the STS-95 life sciences experiments, and Dr. Ronald J. White, associate director of the National Space Biomedical Research Institute in Houston, helped us better understand the science of the mission. Janice Tucker, director of public relations at Muskingum College in New Concord, Ohio, helped us learn about Senator Glenn's hometown. Retired Lt. Gen. Tom Miller, Senator Glenn's very close friend; his wife, Ida Mai Miller; and Rex Hoon, a hometown friend from New Concord, provided personal anecdotes. Jeanette Jenkins of New Philadelphia, Ohio, told us of the senator's early flying days.

Thanks to Chuck Perry, Longstreet Press president; Jill Hasty, a Longstreet associate editor who was the link between Longstreet Press and the *Dayton Daily News*; and the rest of Longstreet's fine staff. Thanks to J. Bradford Tillson, publisher of the *Dayton Daily News*, and the newspaper's senior editing team: Jeffrey C. Bruce, editor; Steve Sidlo, managing editor; John Thomson, deputy managing editor, who helped guide this project; and Ann Hoffman, assistant managing editor. Special thanks to Jeff Adams, *Dayton Daily News* director of photography, who selected the photos you see in this book.

Thanks to NASA, its history office and press office, especially Doug Ward and his staff at Johnson Space Center in Houston. Congratulations to the flight crew of the shuttle Discovery, especially Commander Curt Brown. And of course, thanks to Senator John Glenn, a true American Hero, and his wife, Annie.

We had a wonderful time with this project. We hope you have a better time reading it.

Ray Marcano
Editor, *Back In Orbit*

CONTENTS

THE AUTHORS

Scott Montgomery joined the *Dayton Daily News* as its Washington correspondent in 1996. A native of Long Island, N.Y., Montgomery grew up in south Florida where he was a reporter for the *Daily News'* sister paper, *The Palm Beach Post*. While at the *Post*, Montgomery was on the team that covered Hurricane Andrew and in 1995 was named Cox Newspaper's Writer of the Year, in addition to winning other prizes, for his year-long project about life in a nursing home. Before going to the *Post*, Montgomery worked at a small newspaper on the outskirts of Orlando. He lives in Washington, D.C., with his wife, Christine.

Timothy R. Gaffney is the military affairs/aerospace writer for the *Dayton Daily News* in Ohio. Gaffney is the author of several children's books on aviation and space. He and staff photographer Ty Greenlees received the Aircraft Owners and Pilots Association's 1998 Max Karant journalism award for a *Dayton Daily News* series about their flight around the United States in a small airplane. He is a private pilot and owner of a Grumman AA-1B. He lives in Miamisburg, Ohio, with his wife, Jean, and four children.

36 years later, a lot has changed in the American space program. Glenn rode in the proven high-tech space shuttle instead of being strapped to the top of an experimental manned rocket.

184 ft.

Cramped quarters:

Friendship 7 was a primitive spaceship. The bell-shaped craft had a heatshield at its base and parachute compartment in the top. An escape tower at the top held a small solid-fuel rocket that would pull the capsule away from its booster rocket in an emergency. A small package of solid fuel retro rockets was attached to the heatshield with metal straps.

Inside, the astronaut sat in a form-fitting couch that faced an instrument panel. There was no room to get out of his seat and float around. "We sometimes joked, 'You don't climb into the Mercury spacecraft, you put it on,' " Glenn wrote in *We Seven* (Simon & Schuster, 1962,) a book by the original seven astronauts. Glenn spent nearly four hours strapped into his capsule, including the countdown before his four-hour, 55-minute flight.

escape tower

Bigger, more powerful:

The space shuttle dwarfs Glenn's Mercury-Atlas rocket in size and complexity. The winged orbiter, which carries the crew, is just one element. It mates to a huge external fuel tank that holds about one-half million gallons of liquid hydrogen and oxygen for the orbiter's three main engines. The external fuel tank is flanked by a pair of solid-fuel booster rockets. The solid-fuel rockets drop off after launch and parachute to the ocean, where they're fished out and refurbished. The external tank drops off later and burns up in the atmosphere.

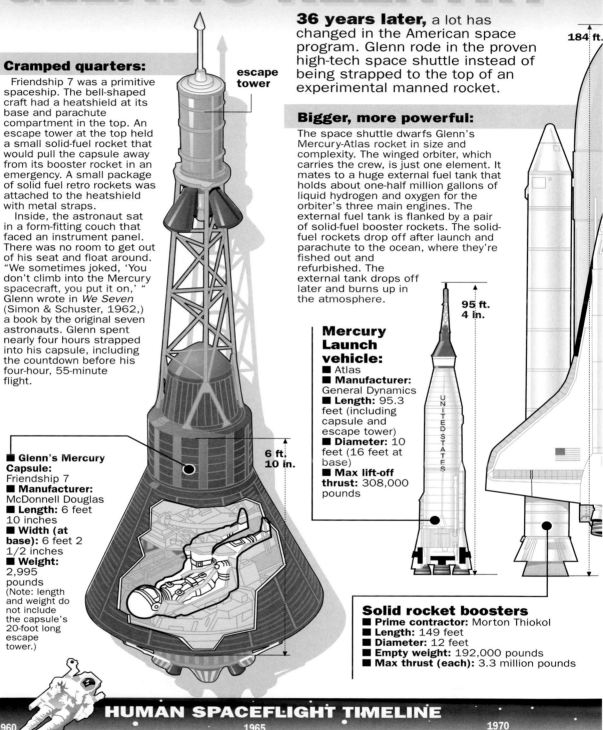

95 ft. 4 in.

Mercury Launch vehicle:
- Atlas
- **Manufacturer:** General Dynamics
- **Length:** 95.3 feet (including capsule and escape tower)
- **Diameter:** 10 feet (16 feet at base)
- **Max lift-off thrust:** 308,000 pounds

UNITED STATES

6 ft. 10 in.

Glenn's Mercury Capsule:
Friendship 7
- **Manufacturer:** McDonnell Douglas
- **Length:** 6 feet 10 inches
- **Width (at base):** 6 feet 2 1/2 inches
- **Weight:** 2,995 pounds
(Note: length and weight do not include the capsule's 20-foot long escape tower.)

Solid rocket boosters
- **Prime contractor:** Morton Thiokol
- **Length:** 149 feet
- **Diameter:** 12 feet
- **Empty weight:** 192,000 pounds
- **Max thrust (each):** 3.3 million pounds

HUMAN SPACEFLIGHT TIMELINE

1960   1965   1970

- **April 12, 1961:** Yuri Gagarin of Soviet Union first human to orbit the earth.
- **May 5, 1961:** Alan Shepard first American in space.
- **Aug. 6, 1961:** Soviet Union's Gherman Titov becomes first in space for more than 24 hours.

- **Feb. 20, 1962:** John Glenn first American to orbit the earth.

- **May 15, 1963:** L. Gordon Cooper becomes first American to spend one day in space.
- **June 16, 1963:** Soviet Union's Valentina Tereshkova becomes first woman in space.

- **March 18, 1965:** Cosmonaut Alexei A. Leonov makes the first spacewalk.
- **June 3, 1965:** Edward White is first American to walk in space.
- **Dec. 15, 1965:** America's Gemini 6 & 7 are the first manned spacecraft to rendezvous in orbit.

- **Jan. 27, 1967:** U.S. astronauts Roger Chaffee, Virgil I. "Gus" Grissom and Edward H. White, scheduled to make the first Apollo space flight, burn to death in their capsule during a ground test.

- **July 16, 1969:** Neil Armstrong and Edward "Buzz" Aldrin walk on the moon.

- **A 197 Uni Saly firs stat

Sources: NASA, Illustrated Encyclopedia of Space Technology, 1998 World Almanac.

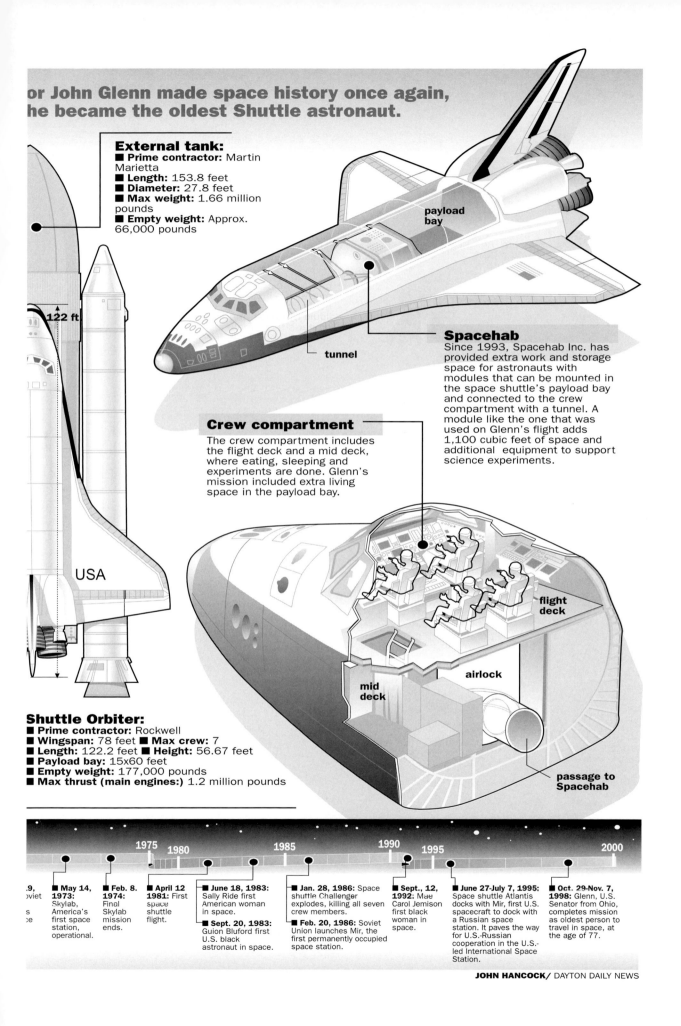

or John Glenn made space history once again,
he became the oldest Shuttle astronaut.

External tank:
- **Prime contractor:** Martin Marietta
- **Length:** 153.8 feet
- **Diameter:** 27.8 feet
- **Max weight:** 1.66 million pounds
- **Empty weight:** Approx. 66,000 pounds

122 ft

payload bay

tunnel

USA

Spacehab

Since 1993, Spacehab Inc. has provided extra work and storage space for astronauts with modules that can be mounted in the space shuttle's payload bay and connected to the crew compartment with a tunnel. A module like the one that was used on Glenn's flight adds 1,100 cubic feet of space and additional equipment to support science experiments.

Crew compartment

The crew compartment includes the flight deck and a mid deck, where eating, sleeping and experiments are done. Glenn's mission included extra living space in the payload bay.

flight deck

mid deck

airlock

passage to Spacehab

Shuttle Orbiter:
- **Prime contractor:** Rockwell
- **Wingspan:** 78 feet ■ **Max crew:** 7
- **Length:** 122.2 feet ■ **Height:** 56.67 feet
- **Payload bay:** 15x60 feet
- **Empty weight:** 177,000 pounds
- **Max thrust (main engines:)** 1.2 million pounds

1975 1980 1985 1990 1995 2000

.9, oviet ce

■ **May 14, 1973:** Skylab, America's first space station, operational.

■ **Feb. 8. 1974:** Final Skylab mission ends.

■ **April 12 1981:** First space shuttle flight.

■ **June 18, 1983:** Sally Ride first American woman in space.

■ **Sept. 20, 1983:** Guion Bluford first U.S. black astronaut in space.

■ **Jan. 28, 1986:** Space shuttle Challenger explodes, killing all seven crew members.

■ **Feb. 20, 1986:** Soviet Union launches Mir, the first permanently occupied space station.

■ **Sept., 12, 1992:** Mae Carol Jemison first black woman in space.

■ **June 27-July 7, 1995:** Space shuttle Atlantis docks with Mir, first U.S. spacecraft to dock with a Russian space station. It paves the way for U.S.-Russian cooperation in the U.S.-led International Space Station.

■ **Oct. 29-Nov. 7, 1998:** Glenn, U.S. Senator from Ohio, completes mission as oldest person to travel in space, at the age of 77.

FOREWORD

By Scott Carpenter

I first met John Glenn in early 1959 during the Mercury selection process at Wright-Patterson Air Force Base in Dayton, Ohio. We may have brushed shoulders before, when we were both stationed at the Patuxent Naval Air Test Center in Maryland, but by the time the Mercury crew all got together, every one of us knew him by reputation because of the transcontinental speed record he held. His subsequent appearances on the television show "Name That Tune" only enhanced his image as the heroic Marine fighter pilot and test pilot. He was a fine aviator who had demonstrated courage under fire and perseverance in all sorts of difficult situations. He stood for all things that Eagle Scouts and America stand for, and he cut a dashing figure doing it. It has been said many times that in John Glenn you could see all the good things like Mother, apple pie, and the American flag. He brought an exemplary image and a valuable talent to the Mercury astronaut group.

After a distinguished career as a fighter pilot in WWII, he returned to the Navy Test Pilot School where he honed his pilot's skills and began probing some new frontiers, both personal and aerodynamic. After we got together in Mercury, I came to realize that he had single-handedly conceived and realized the Project Bullet flight, which would establish the

transcontinental speed record and the first supersonic crossing of the U.S. from coast to coast. That took a good measure of dedication and foresight and hard work, for which I had a great respect. I also was pleasantly surprised to see that even though that flight and the television show that followed had occurred quite some time before we all got together, he still carried around a folder full of fan letters from "Name That Tune" that he had not yet answered. He told me he planned to answer every one of them; I don't know whether he did or not, but I do know that he tried. He had no way of knowing then how many letters lay in wait for him after his orbital flight, but his fairy godmother could have easily and correctly said, "John, you ain't seen nuthin' yet."

We began our training at Langley Air Force Base in Virginia, a little south of Washington, in early 1959. Langley was the home of the Space Task Group, which was the part of NASA that had responsibility for Project Mercury. The training program, which we all had a hand in formulating, was one of the most exciting times any aviator could ask for, and John ate it up. His single-minded dedication to the task at hand was a thing to behold. That dedication to his family and career led John to move alone to Langley and leave his wife, Annie, and children, Lyn and Dave, in Washington so that the kids could stay in the school they had attended for so long. All of us though, including John, missed having Annie and the kids with us in Langley, but there was a lot of travel in those days, and all of us spent more time away from our families than we wanted.

John had his own philosophy about everything he did, and he stuck by it. He never yielded to the need the rest of us had to drive high-performance automobiles. Prior to and during the period when we were driving Cobras, Corvettes, Ferraris, and other muscle cars, John drove back and forth from Langley to his home in Washington in a little car called a Prinz, I think. It got something like seventy miles to the gallon, took forever to get to forty miles per hour, and held two people . . . but it was a convertible, which was its only good feature, to my way of thinking.

During all the long hours, hard work, and time away from home — to say nothing of the rocket explosions at the Cape and the early Soviet successes — we had many good times. Most of those involved making fun of or playing jokes on each other. Everything we did, however, was underlain by an intense but good-natured sense of competition.

One day, John and I were scheduled to catch an early afternoon flight out of Friendship Airport, now called Baltimore-Washington. We were late for some reason or other (not unusual in those days). We were speeding (forty-five miles per hour?) down the divided freeway with the top down when we decided I should get the tickets out of our briefcases to make sure everything was in order and to get them ready to turn in at the gate as we chased after the airplane.

It occurred to me to give John's patience and temper and sense of humor a little test. I let a couple of the useless ticket envelopes fly out of my hand, up into the slip stream and way back behind us and scattered to the winds. I said something like, "Well, there go our tickets." He thought about that for a second or two, realized that there was no place, no time, and no way to turn around, and just slowed down and laughed. He passed the test just fine, but I did notice that the sound of his laughter changed when I told him I had just lost envelopes. We did make the flight.

John was so disappointed, and I was surprised, when Al Shepard was given the first flight. I felt sure that John would go first and so did he. But he got over that soon enough, and the Mercury astronauts became inseparable, all for one team, united in support of Al and the flight of Freedom 7. The camaraderie, esprit de corps, and mutual respect soon got us through that and many other tough times. It pleases me to note that those measures of our fraternity endure to this day.

When John was given the orbital flight, I was assigned as his back-up pilot. I was pleased because I had asked to be his back-up on one of the questionnaires we were given before flight assignments. John and I had begun to work together very well by then.

During the preparation for his flight in Friendship 7, we became very close. We spent nearly every waking moment working together, or at least working on the same tasks, and my respect for the man and his work ethic was enhanced even more.

John was engaged in a task never done before. He was planning to budget a four-and-a-half-hour period of his life to make the best of an opportunity no one had ever had before. He was to test a spacecraft in a flight environment it had never seen before. He was to test himself in a flight environment he had never experienced before. There were many unknowns about both man and machine that he would probe, and when unknowns are probed in any flight regime, danger is always present. John and Annie were both aware of this fact of a test pilot's life. They discussed it openly, decided the benefits outweighed the risks, and both pressed on without dwelling on it further.

The perseverance and diligence that are John's hallmarks paid off handsomely during the final phases of the preparation for the flight of Friendship 7. A number of frustrating delays and holds and scrubs during the last three months before the lift-off seemed to strengthen his resolve to get his flight under way.

Everything went as planned on the flight, with the exception of an indication just prior to retro fire that the heat shield had become detached from the capsule. The problem was to determine if it was sensor failure or if it was really true. If the latter were the case, it represented a serious problem on entry. The safe way out was to believe the indicator and leave the retro pack on after its rockets fired, so that the heat shield straps would help secure the heat shield to the capsule.

The retro pack was incinerated during reentry, but with no adverse impact on the flight except for a more colorful display out the window than John expected.

The flight was a monumental triumph for NASA and the United States. It served notice to the Soviets that we were in this competition for the long haul, and that we were breathing down their necks. It provided a great reward for the tens of thousands of Americans without whom the flight could not have happened. And it created a new and national hero, whose image and charisma made anything he imagined possible.

He has now returned to his roots in spaceflight. It couldn't have happened to a better person.

Scott Carpenter was the second American to orbit the Earth, piloting his Mercury capsule Aurora 7 for three orbits just three months after John Glenn made his flight in 1962. A Colorado native and Navy pilot, Carpenter took a leave from NASA to return to the Navy in 1965 where he participated in an aquanaut project in which he spent 30 days living on the ocean floor.

Carpenter went on to help NASA design the Apollo Lunar Landing Module, and then worked as director of aquanautics for the Navy's Sealab II program. He retired from the Navy in 1969 to found Sea Sciences, Inc. a company involved in improving use of ocean resources. Carpenter lives in Vail, Colorado, and has written two novels about the Navy SEALS.

The space shuttle Discovery would carry six astronaut heroes and one American legend into space.

CHAPTER 1
THE LAUNCH

The countdown clock had been running for days, just another piece of hardware out on a lawn crowded with hardware. Bleachers, cameras, trucks, cars, air conditioners, generators, stage lights, stagehands, stages: they blocked out the grass like a great mechanical picnic. Television trucks cranked their satellite receivers toward heaven like metal umbrellas flipped to catch the rain — on a day when there was no rain. The dishes flashed as they moved under a perfect sun. From this occupied yard at the press site, beyond the giant clock planted in the middle of the grass, the mass of steel and concrete called Launch Complex 39, Pad B was clearly visible.

When the clock started on Monday no one cared. The space shuttle crew came in that day from their training base in Houston and gave the customary ready-to-fly thumbs up, and any eyes diverted from that entrance looked away only long enough to learn the latest about a frightening hurricane swirling in the Caribbean. That day, the National Aeronautics and Space Administration (NASA) put the chances of good weather for the launch at 60 percent, then just a day later bumped it to 100 percent when the

hurricane surged off to the west. Then there was interest in the crew's traditional family barbecue — off in quarantine, invited guests having submitted to a physical exam — and after that the traditional last goodbye out at the foot of Pad B. An enormous steel lattice had been rolled away, revealing the space shuttle Discovery, tipped on its tail and aimed at the sky.

So, all of these prelaunch activities, starring Discovery's seven astronauts, had pushed the ticking countdown timer into the background. Until today. Launch day was here, and the clock had become vitally important. And now, at this moment, T-minus five minutes, it was more than important, it was the center of the universe because it did something nobody expected it to do.

It stopped.

For three solid years John Glenn had been working toward this day, and he had been wishing for it for another thirty-three years before that. The vigil began the moment his searing Mercury capsule splashed into the Atlantic Ocean on a February day in 1962, making him the first American to orbit the Earth. That day also made him a hero in a time when being one meant exemplifying the virtues of courage and self-sacrifice while facing down steep odds in pursuit of a noble cause. In the years since, Glenn left the space

STS-95 space shuttle Discovery astronauts walk out from the astronaut quarters at the Kennedy Space Center en route to the launchpad for John Glenn's historic return to orbit.

program because they wouldn't let him fly again, became a wealthy businessman, and then rose to even greater prominence as a one-time presidential contender from his seat in the United States Senate, where he represented Ohio for twenty-four years. Recently, he had decided to retire from politics, but he never let the flame go out in his desire to return to space. And lately, he had been pressing NASA to let him do it, to let him go back up, this time as a test subject in biological experiments on the effects of weightlessness on an aging man. After lining up doctors and scientists to support his line of study, Glenn finally got NASA's blessing last January. So now, if the clock in the grass would ever get to zero, the grand achievements of a lifetime would more or less double for John Glenn, and in the same instant a magical day in America's history would be reincarnated. If the clock reached zero. An event so long in building and so rich a symbol of how life can be expanded, how life can be added to, in the end owed its fate to a process of subtraction. With this clock, history itself had been put on a schedule measured to the hundredth of a second. A sunny afternoon would turn into another day of consequence in the story of humankind, another day when the human frontier would advance discernibly. At the age of seventy-seven, John Glenn had been found fit enough in mind and body for this; today he would travel into space aboard a rocket ship. He would become the oldest person ever in space by a wide margin.

If the clock reached zero.

"I'm nervous and I'm excited. I feel like a kid at his first Christmas," President Clinton told CNN about an hour before liftoff. "I'm very excited about this."

Even the President of the United States had been drawn to the launch, and his presence escalated the festival-like atmosphere of the day into something bordering on frenzy. The last time a president attended a launch was in 1969, when President Nixon watched Apollo 12 shoot to the moon. In most ways, this space shuttle launch was routine, especially in comparison to a trip to the moon, but still it had a universal appeal that made President Clinton perhaps the highest ranking viewer, but not the most famous. Hollywood stars including Bruce Willis and Leonardo DiCaprio were on hand. Heavyweight boxing champ Evander Holyfield had come. And Spain's King Juan Carlos also was in the stands to see Pedro Duque, a citizen of Spain and a member of the seven-person Discovery flight crew, become his country's first to reach Earth orbit.

"Well, of course there is the John Glenn factor," President Clinton said when asked about his reason for being there. "Senator Glenn is a very good personal friend of Hillary's and mine, as well as an ally, a colleague. And like all Americans, I'm thrilled that he is going up today. But also this really is the last launch before we begin to put the International Space Station up. So John Glenn began this first phase of our space program, and he's ending it just before we start on the space station.

"So it's very exciting. It's important for the space program. But it's a great day for America, a great day for our senior citizens, and I hope that all Americans share the exuberance that I feel today."

All estimations were that they did. Hundreds of thousands had piled onto central Florida's Atlantic coast, lining the causeways with campers, packing the sand on the beaches with a legion of bare feet. Just about every outfit with a television signal was carrying the launch live. And all the while, the crew of Discovery: Commander Curt Brown; Pilot Steve Lindsey; Mission Specialists Dr. Scott Parazynski, Stephen Robinson, and Duque; and Payload Specialists Dr. Chiaki Mukai and John Glenn, were seated in the space shuttle. Their feet were in the air as they lay on their backs in padded metal chairs, living their last breaths of natural air before nine days of bottled oxygen. Water coursed through plastic tubes embedded in their blue long johns, keeping the astronauts cool while wearing 100 pounds of gear packed into their pressure suits. Astronauts refer to this outfit, worn only for launch and reentry, as the "pumpkin suit" for its color, and perhaps also for its bulky contribution to the human form. Liftoff was scheduled for 2 p.m., but that time had come and now it was gone and Discovery was still on Earth. All of the built-in countdown holds, including the one that allows the launch

The countdown stopped at five minutes and nearly stopped again. But Discovery lifted off.

close-out team to batten the shuttle door and flee the coming rocket ignition, had gone by. Five minutes to go, and the clock was frozen for something NASA couldn't control.

It's not uncommon for small airplanes to gather by the dozens in the vicinity of the space center on launch day, just for the thrill of watching it go roaring by. For the biggest of launches, air controllers have seen maybe 100 planes congregate in the sky. But for the launch of Discovery, John Glenn's return to space, 300 or more planes had flown in to be near the site, and as they buzzed around a

couple of them had crossed into protected air space. The countdown clock was on hold while the Federal Aviation Administration dispatched a trio of its own small planes to chase the intruders away.

All of this seemed rather silly as it was explained from the loudspeakers at the space center, and a smattering of satisfied applause followed when at long last the clock began moving again.

But inside NASA's launch room, the problem wasn't solved. The air around the launch site was so thick with planes that the air control radar at nearby Patrick Air Force Base became "saturated with data" as it tried to track so many objects. For more than four minutes, the Air Force couldn't tell how fast the planes were traveling and had to rely on a back-up system routinely run for launches by the FAA out of Miami. With two minutes and forty seconds left on the launch clock, the air traffic controllers called launch control with news it did not want. More planes in the safety zone, put another hold on the launch. This was not a good thing because inside of five minutes, the countdown toward launch becomes more and more inflexible, making holds in those final minutes more and more likely to cause a scrub for the day. Ground crews already had disconnected the nozzle that constantly replenishes the shuttle's liquid oxygen tank, which loses fuel as the ultra cold liquid boils off, and that meant this hold simply could not last more than five minutes and eighteen seconds, period. After that, the liquid oxygen would warm up too much from its minus 297 degrees and change density, harming its efficiency as a rocket fuel. The President, the King, the press, and the world were staring at a countdown clock moving effortlessly toward history, and none of them knew that NASA's own launch director had a sinking feeling that everyone would have to come back tomorrow.

A lingering question among observers of the space program was exactly what the countdown operators at NASA planned to say when liftoff arrived. The first time John Glenn flew, his close friend and back-up pilot, Scott Carpenter, interrupted the countdown to say "Godspeed, John Glenn," seconds before the rocket lit. Those three words instantly crystallized into a lasting piece of America's history in the space program, more famous than Carpenter's own trip around the world three months after Glenn flew. Lately, "Godspeed, John Glenn" had

been resurrected in anticipation of the shuttle launch, along with every other detail of America's first orbital flight. The festival surrounding Discovery's launch had the feel of old home week for many of the big names in space travel. Along with original moon walker Edwin "Buzz" Aldrin wandering the grounds, and newsman Walter Cronkite out of retirement working for CNN, the three surviving Mercury astronauts conducted interviews at will. Notably absent was Al Shepard, who had planned to come for the launch but died of leukemia three months earlier. And NASA itself borrowed heavily from the Mercury theme, stitching "Godspeed, Discovery" on the shuttle processing team's mission shirts. But saying it during liftoff seemed like it might unfairly recognize that one old man riding in the shuttle's windowless mid deck at the expense of an exceptional crew led by Brown, who was making his fifth shuttle flight. The captivating story of a hero's return to space had been all the rage since NASA announced in January that Glenn was going back up, and heaven knows the agency enjoyed the emphasis on its rugged, pioneering past. Back in John Glenn's day — the first time, not this time — NASA represented America in the struggle against communist oppression, which made Glenn an instant free-world hero when his flight struck back at the Soviet Union's surging space program. But liftoff for the space shuttle Discovery, a here and now enterprise if ever there was one, was no place for nostalgia. On launch day every astronaut is risking life and limb to get into space.

But only if the clock reached zero.

Test Director Doug Lyons, who runs the countdown operation and reports to the shuttle launch director, agreed to the air traffic hold. But since they were within those last five minutes, they couldn't stop the clock until it

STS-95, the Shuttle Discovery, rose majestically.

was down to thirty-one seconds. Lyons was more confident than his boss, even if he was just as tense. He thought they'd clear the planes out in time, but he didn't like the idea of getting down so close to ignition and then stopping. No particular reason, it just creates more time for problems to crop up. The less time the shuttle sits on the pad, the better. And it was frustrating to be in the control room, all systems up and working perfectly, but still not able to get the shuttle off the ground because

of something completely beyond control. At least when it's a system problem, there's something to do. The control room people had discussed the intense public interest in this mission, talked about how to stay focused on the tasks at hand, even expected that somewhere, somehow all this attention would get in the way and they'd have to deal with it. But nobody thought it would happen in these very last moments before liftoff.

With eyes on the clock, the control room

waited for the thirty-one second mark to come so they could throw down the final hold. But then air traffic called. Plane's gone; we're all clear.

A huge sigh of relief was released in the control room, and nobody saw it. The countdown clock never stopped falling. Seconds after the moment when the hold was supposed to happen, the NASA loudspeaker reported: "Less than one minute to the historic return of John Glenn to space."

Ten.

Nine.

Eight.

Seven.

Six.

Five.

Four.

Three.

Two.

One.

A plume of steam exploded beneath the shuttle, which in that instant began rising. As it climbed up the tower it became clear through the steam that a glow, bright as the very sun, was pushing it up.

"Liftoff of Discovery with a crew of six astronaut heroes and one American legend," the control room announced. So there was the answer. No Godspeed, all heroes.

A dull growl drifted over the viewing site, growing with the seconds as the shuttle rode a column of sunlight higher in the sky. The growl was getting clearer and louder and thirty seconds later the air three miles from the launch pad was overwhelmed by a hammering thunder that tingled in the ground and rattled the grandstand roof. By now the shuttle was flying at about 1,700 miles per hour.

"C'mon! Go! Go!" someone shouted.

In two minutes the shuttle was all but out of sight, a lighted dot, a star in the daytime sky. At the press site, in the VIP viewing area on Banana Creek, on the control room roof where the President and First

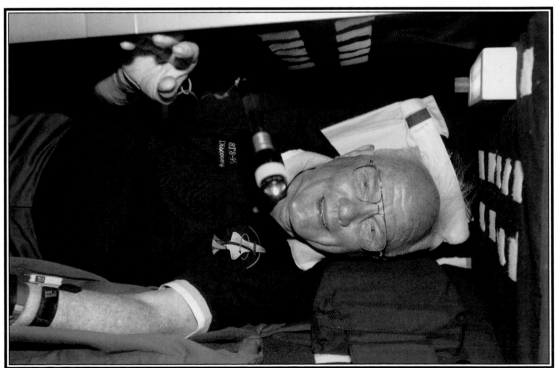

Glenn would be the subject of many experiments in space.

Glenn, aboard Discovery, sips on a grape drink.

it lands.

The first forty-seven space shuttles had landed without the drag chute, so that wasn't a problem. The big danger had occurred the instant the door fell, when this eleven-pound hunk of aluminum, eighteen inches tall by twenty-two inches wide, whacked into the engine nozzle. If it had struck one of the hydrogen fuel lines that runs down the side of the nozzle that could have been disastrous, so NASA's people monitored the engine's performance as the shuttle dragged a thick contrail of smoke up the sky.

Every indication was the engine was fine, crisis averted, and at 2:29 p.m. the countdown voice came back on the loudspeakers. The shuttle had been up for nine minutes, the solid rocket boosters were jettisoned, the fuel tank gone, the engines cutoff. Everything on schedule.

"John Glenn is in orbit for the second time in his life," the voice said.

John Glenn had done it. The view of the world from heaven, which he had not forgotten since that brief taste in 1962, was his again. Using his will, his nerve, his curiosity, and his age, John Glenn had created a day for history out of a sunny Thursday afternoon.

But how?

How indeed. The record books will mark this launch day for the extraordinary feat of a man not unfamiliar with extraordinary feats. But in the context of his life, what's remarkable about this one Thursday is that it comes as no surprise. Looking back, from his boyhood in small town Ohio through his years among the world's most powerful people in Washington, John Glenn's return to space at the age of seventy-seven actually seems logical. Or at the very least fitting. John Glenn specialized in being in the right place at the right time, but not by chance. Over and over, he had a hand in creating the places and the times. When you unfold his

Lady watched, a smattering of applause erupted. People who did not expect to, had tears in their eyes.

But again, in the control room, something else was going on. As the rockets ignited, while Discovery sat on the pad seconds from exploding toward the sky, a white tile from the space ship fluttered to the ground, bouncing off the bell-shaped nozzle of one engine. What was that? With Discovery majestically climbing, engineers in the control room scrambled to analyze video of the falling object. It didn't take long to conclude that it was the door that covers the drag parachute, which is housed in a compartment at the base of the tail and is used to slow the shuttle when life story, it's clear he was headed for history all along.

By the evening of launch day, John Glenn had scurried free of his stowage duties in the mid deck and at last got some of the "window-time" he had promised himself, and everyone who talked to him, he would get on the space shuttle. From 342 miles above the Pacific Ocean, Glenn sounded happy.

"Hello Houston," he said. "This is [Payload Specialist 2] and they got me sprung out of the mid deck for a little while. We're just going by Hawaii and that is absolutely gorgeous."

He sounded like a man who was where he should be.

"First report is great. I don't know what happens on down the line, but today is beautiful and great. . . ."

John Glenn and Annie Castor, his high school sweetheart who eventually became his wife, sit in Glenn's first car, a Chevy coupe. Before "Bud" — as some of his friends called him — became a pilot and astronaut, his need for speed was answered by racing in and around his small-town home of New Concord, OH.

CHAPTER 2
THE NEED FOR SPEED

Swissshhh!

The Cruiser splashed a spray of muddy water as it spun across a wet farm field near the village of New Concord, Ohio. Its narrow tires slid across the slick green weeds and splattered brown mud up the dark maroon sides of the big old Chevy coupe. The redhead was at the wheel, the one with the freckles, and his usual crew of buddies piled in around him.[1]

They called the car the Cruiser, and the driver went by Johnnie, Johnnie G., or sometimes just plain Bud. Anything but John Glenn because John Glenn was his dad, the local plumber who also had a Chevy dealership, which is where this car had come from. The car wasn't much, just a beat-up '29 convertible with two doors and no top that had been languishing on John Glenn Senior's car lot. Senior had given it to Junior, to Bud, the redhead, when he turned sixteen and a half.[2]

Six decades later, the old clunker means as much to Junior as it did back then. Sen. John Glenn, the venerable Democrat from Ohio, smiles and his eyes sparkle when he talks about that Cruiser. His wife, who was his sweetheart back then, blurts a laugh herself. No comment, just a giggle. Glenn's memory quickly conjures the brick streets of New Concord and the primitive machine that gave

him his first teasing hints of power and speed.

"[The] thing wasn't worth more than about fifty bucks, and in fact it wouldn't run when he gave it to me," Glenn recalled. "And the top was long since rotted away. I just carried a canvas shelter half in there, and when I stopped, why, I'd throw that back up. And when it rained, why, it rained, and the water would collect on the floor. So I bored a hole in the floor and let the water run out. That was it."

It wasn't luxurious, it wasn't fast, and it wasn't all that dependable.

"The thing would only run about forty miles an hour wide open, so you couldn't get in too much trouble with it. It was mainly to get around New Concord, Zanesville, or Cambridge once in a while. And it had such big leaks out of the water system that in the winter time I'd just run it dry. . . . If you ever ran it too far, why, it'd get to where the grease on the top of it would just start popping," Glenn said.

No, it wasn't much, but it was his, and that was something at a time when America was still climbing out of the Great Depression and teenagers didn't think of cars as an entitlement. The Cruiser gave Glenn mobility. His house sat on a hill overlooking the big highway — U.S. Route 40, the National Highway, a two-lane brick road that snaked across the hills of eastern Ohio and connected New

Concord with Cambridge and Zanesville, two cities that bracket it on the east and west. In his circle of friends, Johnnie became the One with the Wheels. And, ratty as it was, it gave him a carriage to court his girl, Annie Castor.

But it was even more than that. Sure, it was no hot rod, but its old straight-six engine gave him his first sense of power when he gripped the wheel and his first sense of speed when he mashed the gas pedal. Whether he was thumping down Route 40 or doing doughnuts on a muddy field, Glenn might as well have been Roscoe Turner, standing his powerful Geebee on one wing to round a pylon in the National Air Races.

By all accounts, Glenn's upbringing was extraordinarily straitlaced. When he became the first American to orbit the Earth, an adulatory press made him sound like someone right out of a Norman Rockwell painting: a freckle-faced, small-town boy who lettered in high school sports, sang tenor in the Glee Club, attended the local Presbyterian college and married his childhood sweetheart — just in time to go off to war, no less.

Glenn was such a straight arrow that after his dramatic Mercury flight, reporters who covered his political career found him downright boring.

Born in Cambridge on July 18, 1921, he soon moved with his family to nearby New Concord, a small village of about 2,000 nestled in gentle hills covered with woods and pastures. Its most prominent feature was Muskingum College, a Presbyterian liberal arts school with a small campus of red brick and white frame buildings. New Concord was — and, amazingly, still is — a small, reasonably prosperous community steeped in conservative values. Glenn's Scottish father and Irish mother raised him to believe that virtue and hard work would be rewarded. He never doubted it, and those core values shaped his life.

But so did machines. Tools and machinery were a part of Glenn's environment. His father was a plumber, and he eventually owned a plumbing and heating business in New Concord as well as the Chevy dealership there. He knew the practical value of mechanical skills; when he gave John Junior the old Chevy, the real gift wasn't

Glenn's boyhood home in New Concord is on Friendship Avenue. The street was renamed after Glenn became the first American to orbit the Earth in his Friendship 7 space capsule in 1962.

TY GREENLEES

mobility, but a chance to tear into a car and learn what made it go. "He gave it to me with the idea I'd work on it and learn something about it," Glenn said. "Which I did."

His fascination with airplanes started even earlier. His favorite toys were small metal airplanes. On trips in the family car, he would hold one out of the window to watch its propeller spin in the airstream. When scarlet fever swept through the village one winter, his parents confined him to the house for several weeks. He busied himself building model airplanes out of balsa wood and tissue paper, and soon he had a room full of them. A family trip to Columbus had to include a stop at the airport so he could watch airplanes take off and land. As junior class president at New Concord High School, he picked aviation as the theme for the Junior-Senior Banquet. The cover of the program showed a Ford trimotor — an early airliner — with the inscription, "NCSH Senior Airways, Inc., Dependable Safe Service. Port NCHS to Port Brown Chapel." It described a route from the high school to Brown Chapel, one of the main buildings on the Muskingum campus.[3]

He didn't have wings yet, but he had the Cruiser, and he drove it for all it was worth. Local legend has it he particularly enjoyed the challenge of a one-lane bridge where Liberty Street crossed the railroad tracks. The bridge is a wooden structure that angles up sharply at each end, making it impossible to see if another car is coming across. When cars approach it — yes, it's still there — they generally slow to a crawl and give a loud honk on the horn before creeping up onto the boards.

John Glenn Jr., here at about age five, was the son of a plumber and Chevy dealership owner, John Glenn Sr.

THE GLENN COLLECTION

Glenn's method, townsfolk say, was to fly up the ramp and charge across.

Local legend also has it he once drove the Cruiser into a lake at Muskingum College. That's one Glenn disputes. "They bring that up every time I'm introduced around New Concord," he said with good-natured exasperation. "I don't know where on earth that ever came from, because it never happened."

Fortunately for Glenn, he learned early to rein in his need for speed. Unfortunately, he learned it the hard way. When doing tricks that day on that muddy field — whump! — he spun the Cruiser into a log lurking in the weeds. It dented the oil pan, and when the Cruiser

Journey to Glouster Friday

Top Row—Handschy, Atkinson, Aitken, Hadden, Chess.
Bottom Row—McKinley, Glenn, Wilson, Patton, Hoon.

New Concord High School newspaper with a photo of John Glenn and his basketball teammates.

climbed back onto the roadway, mud wasn't the only thing dripping from it. The Cruiser's oil bled away, and its bearings burned up on the drive home. John Senior confiscated Johnnie G's car keys for a month.[4]

Glenn's Cruiser may be the stuff of local legend because there isn't a lot to talk about in New Concord. Next to Muskingum College, Glenn is the biggest thing the village has ever seen. Its high school is named after him. Ohio Route 83, the only exit to New Concord from Interstate 70, is named John Glenn Memorial Highway, and another road in town is named Friendship Avenue, after Glenn's Mercury capsule, "Friendship 7." The college named its gymnasium after him, and a large scale model of his capsule stands in a campus building lobby.

Local folks say New Concord has changed a lot from the days of Glenn's youth. They point out I-70, which skirts the south edge of town parallel to Route 40, and a manufacturing plant recently built near the highway. The interstate has also drawn ex-urbanites attracted to the pastoral setting.

Population has swelled around New Concord. But an outsider is likely to be impressed by how little things have changed.

The Glenn family home still sits where his parents moved it decades ago, though a local group is interested in acquiring it for a John Glenn museum and moving it downtown. New buildings on the college campus don't alter Muskingum's appearance; its administration building, a stately edifice of red brick with white trim, still overlooks the town from a hilltop perch. The neighborhoods seem intact, and the white frame churches where the Glenns and Castors worshiped still poke their steeples above the housetops. There is a sense of steadfastness in character — neighborhoods of middle-class homes that have been well kept, yards trimmed, their streets quiet and shady. New Concord has changed over the years, but it hasn't let change roll over it.

That's true of Glenn as well. He's still the boy from New Concord. His first car was a convertible, and he's had one ever since. In his final year in the Senate, the elder statesman drove a dark green ragtop Chrysler Sebring. His first girlfriend was Annie, and she's still his girl. The overachiever who strapped into a space rocket three months after his seventy-seventh birthday is the same kid who lettered four times in football and three times each in basketball and tennis, sang in the glee club and the A Cappella choir, played the trumpet in band and orchestra, and played the lead in the senior class play.[5]

Nobody in New Concord says Glenn was a genius or a natural athlete. "He wasn't what you'd call a great athlete. He was kinda slow," recalled Rex Hoon, a retired high school coach who grew up with Glenn. But Hoon said Glenn was a natural leader. "He was one of the regular guys, but he showed this leadership. He was serious-minded."

"A lot of girls liked John, but he was dating Annie," Hoon said.

Glenn was so popular in the school, he even was featured in an undated edition of the New Concord High School newsletter, the *M. & W.*

Newslite. A poetic column on the school's football team included this stanza on Glenn:

A 3 letter man - an athlete
Red hair, green eyes and appearance neat
In climbing life's long happy hill
He'll ne'er need pay the dentist bill.

To this day, it's hard to describe Glenn without talking about his wife. After his Mercury flight, he introduced her to a joint meeting of Congress as "the real rock in my family," and he wasn't exaggerating.[6] She endured long separations during his military and astronaut careers and stayed close by his side throughout his years in the Senate. (He insists he got her approval before signing up for the Discovery mission.)

Such a life is never easy, but hers was all the harder because of a severe, lifelong stuttering problem that made it difficult to hold a conversation and impossible to be a public speaker. "There were always public events, always crowds, and always well-meaning people who went out of their way just for a chance to talk with us. . . . Out of sheer frustration I vowed that someday, somehow, I would be able to give a speech for him — but I couldn't see any way to do it," she said in 1986.[7]

She didn't make headlines or get a ticker tape parade, but Annie Glenn attacked stuttering with the same tenacity her husband displayed. When one therapy program after another failed her, she kept trying, and in 1973, she went through an intensive program that helped her conquer the problem. Now she not only gives speeches, but often handles the radio chores, exchanging critical information with air traffic controllers when she flies with her husband in their airplane.[8]

Like his relationship with Annie, Glenn's future in New Concord seemed as clear as a course line on a chart — plotted out, as that junior-senior banquet program suggested,

JOHN GLENN

Glenn, in 1942, was a student at Muskingum College and studied chemical engineering. He was also learning to fly.

from the local high school to the local college. He enrolled at Muskingum to study chemical engineering after graduating from high school in 1939. Annie, a year older, was already working on a music degree.[9]

But the world was in ferment, and in hindsight, it wasn't a good time for young men or their girlfriends to make lifelong plans. Japan had invaded China in 1937, and in 1939 Nazi Germany began rolling over Europe. War was creeping across the globe.

Even New Concord wasn't isolated. The U.S. government was already building up industrial output as it turned out weapons for beleaguered allies. The Battle of Britain showed how important air power had become, and aircraft manufacturers were scrambling to develop better fighters, bombers, and transports. But if the U.S. itself came under attack, the best planes would be useless without pilots. In 1939, the Civil Aeronautics Authority launched the Civilian

Glenn, #75, on the 1942 Muskingum College football team, was known for his hard work and serious approach to sports.

Pilot Training Program in an effort to build a pool of potential military pilots. It certified colleges and universities across the U.S. to open flying schools for physics course credits. Muskingum was one of them.

The college-directed program was at New Philadelphia Airport, a grassfield air strip with a single, brick hangar built not long before with New Deal cash. The airport was about fifty miles northeast of New Concord, and the college offered daily roundtrip transportation. That seemed like a swell deal to Johnnie G. and four of his buddies; they all wanted to sign up.

But Glenn's parents didn't share their son's enthusiasm. In 1939, flying still looked to them like a dangerous adventure. The program's director, Muskingum physics professor Paul Martin, went to their home and persuaded them that aviation was not only safe, but poised to become the next great growth industry.[10]

The elder Glenns relented, and Johnnie prepared to fly. The college provided the boys a brown Chevy station wagon for the commute to New Philadelphia, and twice a week

they drove out to the airfield, coming home long after sundown.

First it was just classroom instruction, chalkboard diagrams, and textbook lessons on the physics of lift, the elements of aerodynamics. Later they got to the air work. The lessons were given in a Taylorcraft BL-65, an unpretentious two-seater whose shell of fabric made clear that it was a not-so-distant relative of the muslin-covered biplanes of Wilbur and Orville Wright. High-winged and powered with sixty-five horsepower, the Taylorcraft was simple and perfect for teaching the basics of aviation.

Later, Glenn professed love at first flight. "I was sold on flying as soon as I had a taste of it," he wrote in *We Seven*, a book by the original seven astronauts.[11]

And while he soon was entertaining thoughts of some kind of career in aviation, it was not without misgiving. On his application for the basic flying course, Glenn was asked if he intended to follow up with advanced training. "Possibly," he wrote on the page. "Not sure." The date on the application read Feb. 20, 1941. Twenty-one years later to the day — Feb. 20, 1962 — Glenn would pilot a nickel alloy capsule around the Earth as the first American to fly into orbit.

With the Taylorcraft, teenager Glenn mastered the basic maneuvers and soon droned across eastern Ohio on cross-country flights. In flight records from his first lessons, an instructor wrote that Glenn was "eager to learn, relaxed, alert and [showed] good coordination." Harry Clever, who as Glenn's first flight instructor wrote those words, ran the Civilian Pilot Training Program and was a founder of the airfield back in 1927, according to Jeanette Jenkins. Jenkins learned to fly at the same time Glenn did and later worked for Clever.

Today New Philadelphia's airport is called Harry Clever Field, an honor noted in a tall, narrow granite monument that stands alongside an unmarked, weatherbeaten rocket. Back then, Glenn concluded his flight training with a second instructor, Wallace E. Spotts, who like Clever was impressed with the redhead's plane handling and recommended him for his pilot's license.

Glenn got the license and kept driving back to New Philadelphia, or once or twice he rode over to another grass airfield in Zanesville. "Oh, I'd save a few bucks and go rent a plane and go fly for an hour or two," Glenn said. "I think I had a total of like, maybe, sixty hours grand total in the Taylorcraft. . . ."[12]

By his junior year in college, Glenn was getting serious about a flying career. But first, he planned to graduate from Muskingum and marry Annie. Certainly Annie was on his mind, maybe marriage too, that Sunday in December when he drove up the hill to Brown Chapel for her senior recital on the chapel's pipe organ. Winter had started to make a show of itself, and an icy breeze kicked brown leaves across the campus ground. It was the kind of day that could whip a fellow's ears into an aching redness after just a few

ANNA MARGARET CASTOR

MUSKINGUM COLLEGE

Anna "Annie" Margaret Castor was Glenn's sweetheart since high school. She went to Muskingum College to study music.

miles in a convertible with no top, but that's not what made it one of the most memorable drives in his life.

"Just as I was turning up the hill on College Drive, the program I was listening to on my car radio was interrupted by a special bulletin," Glenn recalled in 1997. "I didn't tell Annie about the bulletin until after her recital, but the news that day completely changed the direction of our lives. The day was December 7, 1941. Japan had attacked our fleet at Pearl Harbor. America was at war."[13]

While Annie played "Be Still My Soul," Glenn sat in the chapel — thoughts about war, Annie, and his future racing through his mind. They talked it over that evening, but there couldn't have been much doubt about what he felt compelled to do.

A few days later, Glenn and his buddies took the Army Air Corps physical, signed up, and were sworn in. They were ready to fly and fight for their country. But their country

THE GLENN COLLECTION

John and Annie Glenn on their wedding day, April 6, 1943. Glenn came home to marry Annie after completing Naval flight training just before his first assignment as a flyer during World War II.

— or at least the Army — didn't seem ready for them. They waited for orders to report for training, but none came. They called to check on the hold-up and were told to wait some more. Impatient to get in the action, Glenn and his crew went to a Navy recruiter in Columbus and signed up all over again. The Navy sent orders immediately, and off they went. Glenn was shipped to the University of Iowa in Iowa City for pre-flight training, then to the Naval Reserve Air Base in Olathe, Kansas, for primary training. He finished up his advanced training at the Naval Air Training Center in Corpus Christi, Texas. There he learned he could switch to the Marine Corps. Gung-ho as always, Glenn became a flying Leatherneck.

Glenn pinned on his lieutenant's bars and wings on March 31, 1943. He stopped back in New Concord just long enough to marry Annie on April 6.

Then he was off to war.

NOTES

1) Mercury Seven Astronauts, *We Seven* (New York: Simon & Schuster, 1962), pp. 33-34.

2) Ibid., and Philip N. Pierce, *John H. Glenn—Astronaut* (New York: Franklin Watts, 1962), p. 13.

3) *We Seven*, p. 34.

4) Ibid., pp. 33-34.

5) Muskingum College News Service, biographical sketch of John H. Glenn Jr., January 1972.

6) John Glenn, message to Joint Meeting of Congress, February 26, 1962, reprinted in Pierce, *John H. Glenn—Astronaut*, pp. 178-184.

7) Remarks of Annie Glenn, Muskingum College sesquicentennial, September 24, 1986.

8) Muskingum College News Service, biographical sketch of Annie Glenn, undated.

9) Glenn's enrollment in Muskingum reflected his persistence and loyalty. Although he quit college his junior year to join the Marines, and managed to become a test pilot and astronaut without a college degree, he still wanted to get that degree. In 1962, college officials reviewed the coursework he had taken as part of his military career and decided it qualified him for a B.S. It was awarded to him at the June commencement. Nevermind that a year earlier, the same college had awarded him an honorary doctorate! Neither Glenn has forgotten their alma mater; as of 1998, they were both still serving on Muskingum's board of trustees.

10) *We Seven*, p. 35.

11) Ibid.

12) Interview with U.S. Sen. John Glenn, Washington, D.C., May 21, 1998.

13) Remarks of U.S. Sen. John Glenn, Muskingum College, February 20, 1997.

John Glenn was a Naval air cadet in 1942 and a year later was fighting in World War II.

CHAPTER 3
THE WARRIOR

"Funny how the bullets sparkle when they hit a plane like that. Just light up like little lights every time a bullet hits."

Diving through the sky, aiming at the tiny MiG-15 far below, hands gripping stick and throttle, sweat running, closing rapidly now, pulling the trigger, feeling the shudder of six machine guns firing, watching the bullets rush the hapless MiG and flash against its metal skin.

John Glenn — Johnnie G., redheaded Bud, the freckle-faced Glee Club singer, the Presbyterian boy who married the girl next door — was a warrior. He was hammering that MiG, blasting it, dooming it with the squeeze of a finger.

"Really had them pouring into it. Just chopped it up good, until it finally flamed," he wrote in a letter to his family.

It was July 12, 1953. After Navy flight training, Glenn had switched to the Marine Corps, and joined a fighter squadron. He had already been through one war, World War II, and had flown sixty-three missions in Korea. Yet in all those past missions, his targets had always been on the ground, not in the sky, and despite the danger of dodging the Earth and enemy fire at the same time, he had not yet experienced the knightly air-to-air combat for which fighter pilots lust. Until now. Now the Marine pilot was flying an F-86 with an

Air Force unit under a temporary exchange program he had volunteered for, and he had just shot down his first enemy aircraft. Not just any aircraft, but a jet at that, a Russian-built MiG-15, a technological match for the F-86. This was the Jet Age, a hot moment in the Cold War, and Glenn was reveling in it.

"Of course, I'm not excited about it at this point. Not much!" he wrote.[1]

No, not much. Glenn posed for a photo standing in the cockpit of his jet with the words "MiG Mad Marine" painted on its side. (Legend has it the slogan reflected his adrenaline-pumped elation, but Glenn himself says some Corps buddies put the slogan on the F-86 to poke fun at him for flying with the Air Force — just as they painted "Join the Marines" across the wings on the bottom.)

Glenn had come a long way from the Navy flying cadet in Corpus Christi, Texas, who signed up for advanced flight training in multi-engine airplanes so he could make a living with an airline after the fighting stopped. He had faced death time after time and won a fistful of air medals and Distinguished Flying Crosses. Still, "MiG Mad Marine" hardly described the man. What continued to define Glenn was the same single-minded determination he brought to bear on all of his goals — whether it was marrying his childhood sweetheart, leveraging flight training more suited to an airline career into a fighter assignment, or, decades later, wangling a ride on a space shuttle.

Retired Marine Corps Lt. Gen. Thomas H. Miller remembers Glenn's war years. He went through many of them with Glenn. They met when they were Navy cadets, and in a long series of assignments to the same or nearby units, they grew lifelong ties. Miller, in fact, still lives in the house in Arlington, Virginia, that was next door to the Glenns in 1962 — the house where Annie and the Glenn children gathered to watch Glenn blast into space.

"We'd both built a lot of model airplanes. I lived in a small town in Texas, and he lived in a small town in Ohio. He had two years of college, he was completing his sophomore year. . . . I was a junior at the University of Texas," Miller recalled.

They met through a mutual friend after Glenn was transferred to Corpus Christi. They spent a lot of time together, talking about their lack of college degrees and their prospects for flying careers after the war. Both wanted to be aircraft carrier pilots, flying off and onto the heaving, rolling decks of ships at sea, the ultimate challenge for a Navy flier. But there was little call for such skills in the commercial aviation world.

"Although we wanted to fly off the carrier, we thought we'd better forego that and take the multi-engine and the sea plane training so we'd have an occupational skill when we got out [of the service]," Miller said. "And we'd count on somehow managing to twist ourselves into an attack squadron once we got graduated and got our wings."

Pragmatically, they picked multi-engine training over the allure of fighters. They took their advanced training in PBYs — ponderous, multi-engine seaplanes they flew off the water of Corpus Christi bay. They were already thinking about what they would do after the war, and neither saw themselves spending their careers in uniform.

But a minor event changed both their lives. One day after lunch, they spotted a notice on the barracks bulletin board. It listed the top ground and flight students who were eligible to switch out of the Navy and receive commissions in the Marine Corps. Glenn and Miller were near the top of the list. A Marine captain just back from Guadalcanal was going to talk about it that evening.

"Our immediate response was, 'Oh, hell, we don't want to be a damn Marine,'" Miller said. But they had nothing else to do that evening, so they decided to attend the briefing. The officer wasn't even an aviator, but Miller said he was "a super salesman," who painted a vivid picture of the fierce jungle battle that began as a fight for an airfield on the largest of the Solomon Islands. And this captain played up the value of air support in one of America's first Pacific Theater victories. It was a thrilling account.

"Well, to make a long story short," Miller said, "we became Marines."

They reported to the Marine Corps air station at Cherry Point, North Carolina, where they hoped to fly Douglas Havoc bombers. But the place was swarming with young officers. Glenn received lieutenant's bars for the Marine Corps reserve on March 31, 1943. Glenn and Miller were packed off to Camp Kearney Mesa near San Diego, California, and an unglamorous assignment to a transport squadron.

Taunting them from across the airfield was a squadron of hot new Grumman F4F Wildcat fighters. This time, Glenn's determination went into high gear. He decided he and Miller would get to know the fighter squadron's commander and find some way to get transferred into his unit.

The major was a young officer himself, Maj. J. P. "Pete" Haines. Glenn and Miller intercepted him in the mess hall. Their enthusiasm and determination won him over, but

Haines pointed out they would have to go up through their own ranks to get the transfer approved. Following military protocol, Miller dutifully went to his squadron commander and got approval to solicit the next level, the group commander. Not Glenn. Glenn went straight to the lieutenant colonel in charge of the group, who thought everything was fine — until the squadron commander learned Glenn had gone over his head. He complained to the colonel, who taught Glenn a lesson by cancelling the transfers for both him and his buddy.

Glenn's gung-ho gaffe cost them several months, but they finally won approval to transfer to Haines' fighter squadron. In June 1943, they flew from Camp Kearney to El Centro, where temperatures above 100 degrees seared them as they stepped out of the transport plane. "It was the hottest place this side of hell," Miller recalled.

So were their airplanes, as it turned out. They soon traded in their Wildcats for the Chance Vought F-4U Corsair — the big, powerful, inverted gull-wing fighter-bomber whose speed, performance, and payload capacity made it one of the legends of its time.

To hear Glenn tell it, flying fighters was about as romantic as math class. "I took flying very seriously, and I did it very professionally. [If] you're a good fighter pilot, you're also a very studious pilot. You're a very thoughtful pilot," he said.

Miller recalls things a little more colorfully. They trained intensively, he said, but they also couldn't resist the adrenaline rush of flying fast and low in tight formation, exploring the limits of their planes and their fledgling skills, play-fighting like lion cubs. Miller said they used to hurtle side-by-side down a big irrigation canal along the California-Mexico border, actually flying below the banks of the canal. Walking up, you wouldn't have seen them race by. "We used to fly in and tap wings, bump each other's wings," he recalled.

And they would chase each other in loops, loop after loop after loop, like squirrels chasing each other up and down a tree trunk. "One night we went out flying together and he got doing that at night, and I don't even remember how many [loops], but it was much more than I would have cared to do," Miller said.

Glenn, now the sedate senator, doesn't like to talk about these things. When confronted with Miller's recollections of Glenn's "serious," "professional" approach to flying, he groaned good-naturedly. "We're in trouble now," he said with a broad smile. "I deny everything." But then he conceded, vaguely, that he might have scratched that itch for speed once, or maybe twice. "You've got to put it in the proper time frame," he reasoned. "We used to do things in airplanes, that if you did them now you'd probably get locked up. But people sort of accepted it back in those days."

After all, they were at war. But the war seemed to be eluding them. While Glenn and Miller looped endlessly over El Centro, American forces landed in the Marshall Islands. When their unit, Marine Fighter Squadron 155, was finally transferred overseas in February 1944, they found themselves in Honolulu. Then they spent four months with a defense wing, guarding Midway Island's submarine base against attacks that never came.

They finally saw combat in the Marshall Islands. But instead of using their powerful Corsairs to shoot down Japanese Zeros, Glenn's unit was assigned the dangerous and unglamorous task of strafing and dive-bombing atolls — ring-shaped islands formed by coral reefs. The battles for the islands were so costly that U.S. commanders decided to bypass many of the smaller ones and simply bomb Japanese forces there into submission — or oblivion. In June 1944, Squadron 155 was sent to Majuro Atoll in the southeastern Marshalls.

Their first strike was against Taroa Island in the Maloelap Atoll, about 150 miles north of Majuro. They launched their Corsairs across the blue Pacific — droning north, the flat coral atoll of Majuro disappearing behind them like a smoke ring — watching for Maloelap to materialize out of the blue vastness ahead. It would be a sobering mission for them all, but especially for Glenn.

Their job was to strafe the island just ahead of a squadron of slower dive bombers, to suppress anti-aircraft fire against the bombers. "We'd go in for a fighter sweep while the bombers were rolling in and starting their dive," Miller said.

Glenn and Miller each led a wingman. They rolled into their strafing run, diving fast to startle and elude the gunners on the ground. Then they pulled out, Miller said, but "John's wingman never showed up." Glenn's wingman, Lt. Miles "Monty" Goodman, had simply disappeared. "John and I went back, and the shore batteries were shooting at us," Miller recalled. But all they could find was the yellow stain of Goodman's dye marker. Miller thinks Goodman was hit coming in and never pulled out of his dive. The marker must have been triggered in the crash, because nothing else remained. "I think it had a great maturing effect on both of us," Miller said.

Goodman was no anonymous flyer; he and Glenn and Miller had started growing tight in the unit. With Glenn's own fondness for music, he hit it off with Goodman, who did a decent Sinatra, complete with a mopstick for a microphone. One night the three of them had gotten busted for singing in the shower long after midnight. The skipper had to come in and break it up so people could sleep. And then, on combat mission number one, Goodman died. In *Glenn — The Astronaut Who Would Be President*, author Frank Van Riper reported Glenn "cried

unashamedly" after he landed.[2]

But the missions went on and on. Glenn and the other pilots spent hour after hour flying their single-engine planes over the Central Pacific, hours of monotony interrupted by moments of furious diving and firing and jinking to avoid streams of enemy fire. Sometimes the sky itself became the enemy: tropical storms would appear without warning, bucking the planes with violent winds and blinding the pilots with clouds and rain.

They were fighter pilots, and they chafed at the lack of Japanese Zeros in the sky. Late one night, Glenn and Miller were rousted from their sleep and ordered to go after Japanese fighters that had been reported at an atoll 250 miles away. They had to launch their Corsairs at 2 a.m. to reach the atoll at first light. The flight of four — Haines and his wingman as one pair and Glenn and Miller as the other — swept around the atoll, hunting for fighters. "We were trying, just hoping we could find a damn Japanese airplane we could shoot up. But, hell we couldn't find any," Miller said.

All they found was a seaplane base with already shot-up airplanes parked on it. "John and I went in and cleaned those out pretty well," Miller recalled with a chuckle.

While they found no opportunities to dogfight, they worked at making their ground attack missions more effective. Soon the Corsairs were lugging 1,000-pound bombs to their targets. Glenn's squadron experimented with ways to improve their accuracy — going steeper, slower, and lower, even to the point of putting down their landing gear in dives to control their speed. "We carried some of the heaviest bomb loads any squadron had ever done. And did it on a routine basis, and other squadrons copied what we were doing," Glenn said.

Glenn's squadron also developed bombing tactics for a new kind of bomb — the

DEFENSE DEPARTMENT PHOTO

Glenn flew sixty-three combat missions during the Korean War, including one in which anti-aircraft fire ripped hundreds of holes into his Panther jet.

napalm bomb. Napalm is an inflammable, jelly-like substance that sticks to whatever it hits and burns furiously. Glenn's squadron attacked Japanese targets with napalm bombs throughout the fall of 1944, improving their methods with each strike.

The fiercest raid came on November 12 with an attack on a Japanese base on Jaluit Atoll. Japanese forces had been quietly rebuilding their strength on the base, and previous bombing attacks had not eliminated them. Glenn's squadron and two others were ordered to burn them out. The Corsairs were loaded with napalm bombs. Glenn's squadron carried the heaviest loads — three 1,100-pound bombs of bottled hell. To carry the weight, the Corsairs were stripped of their

guns and ammunition.

Glenn led one flight, and they rolled into steep dives. Anti-aircraft fire streamed past them. Releasing their bombs at the last moment, the Marines peeled away as the island erupted in angry fireballs. Flames washed over the base like a molten tsunami and burned for hours, finally obliterating thirty-five of the base's sixty buildings. Glenn led a similar raid a week later. The attacks earned him an air medal — the first of eighteen he would earn in two wars.[3]

Glenn ended his overseas tour of duty in February 1945. U.S. forces were securing the Pacific by then and attacking the home islands with an unrelenting bombing campaign. Two atomic bombs on two cities —

Glenn had come a long way from the Navy flying cadet in Corpus Christi, Texas, who signed up for advanced flight training in multi-engine airplanes.

U.S. MARINES

Hiroshima on August 6 and Nagasaki on August 9 — forced Japan to agree to surrender on August 14, with the formal surrender on September 2.

By then, Glenn was wearing captain's bars. Both he and Miller had decided to become career Marine Corps officers. Miller said it was more like an emotional bonding than a logical career choice. "We lived together under extreme hardships and extreme dangers in the war, and the more you do this the more you become a family," he said. "And the longer you stay in, the more difficult it is to get out."

By the end of the year, Glenn had also become a father — Annie delivered their son John David on December 13, 1945.

At the end of the war, Miller was assigned

to a fighter squadron at Cherry Point. Glenn was there briefly, then he was transferred to Patuxent River Naval Air Station in Maryland to wring out new airplanes as a service test pilot.

Glenn's job was to fly new airplanes hour after hour, building time on them rapidly to uncover any maintenance problems that might crop up later in combat. Miller says Glenn and other test pilots flew the planes twenty-four hours a day in three eight-hour shifts.

Miller never needed to set his alarm clock when Glenn had the midnight-to-eight shift. "Gosh, six or seven times in a couple of months, we'd wake up in the morning, six o'clock, with an airplane just about going right through our house." It was Glenn, giving his buddy a wake-up call — blasting Miller and his wife Ida Mai from bed with the full-throttle snarl of an F8F Bearcat. It was a friendly annoyance Miller still relishes. "I'd usually go out in my shorts and stand out in the yard and wave a flag at him and he'd wiggle his wings. . . . We were living out in the country, so he didn't have any problem. Of course, if you tried to do that today, you'd get court martialed so quick. . . . "

Miller was transferred to Pax River just as Glenn was sent down to Cherry Point. Soon, Glenn was overseas on another tour, this time to China and then to Guam in the Marianas Islands. He arrived there early in 1947.

Glenn had been there just a few weeks when he received a message that Annie was in serious condition in a hospital in Zanesville, a town not far from New Concord in Ohio. She had given birth on March 15 to their daughter Lyn. Everything had seemed fine, but an infection set in, and Annie's temperature soared to a life-threatening 106 degrees. Glenn rushed home to find his wife delirious with fever, but ice packs and a relatively new wonder drug called penicillin slowly brought her back from danger. Glenn returned to

Guam, but he was allowed to send for his family. Annie and their two children joined him, and they turned a metal quonset hut into their island home.[4]

Glenn rotated stateside again in 1949, serving as a flight instructor with an advanced training unit of the Naval Training Command at Corpus Christi. He also found time to get checked out in jet aircraft, logging more than seventy-five hours of jet time before being sent to the Marine Corps' amphibious warfare school in Quantico, Virginia.[5] There, he and Miller crossed paths again. Miller was an instructor in the air maintenance school, and since they both had desk jobs, they had to get innovative to keep up their flying time — for the extra pay and also to meet the minimums required to maintain their pilot proficiency. The answer was to duck out at night and fly big transport planes out all over the country. Around holiday time, "much to the disgust of our wives," Miller said, he and Glenn would load up young Marines and deliver them to assorted destinations for their family leave time.

By February 1953, Glenn was back in another war — the Korean Conflict, the so-called United Nations' "police action" that became the first war of the jet age.

Glenn flew the Grumman F9F Panther with Marine Fighter Squadron 311 based at Pohang Dong. Its armament included guns, rockets, and bombs. Again, Glenn found himself in a ground attack role. His drive for accuracy, flying in close to make sure he hit the target, kept the ground crews busy patching up his planes. After one mission, his crew chief counted 203 holes in Glenn's plane.

Glenn's formal work as an experimental test pilot was still in the future, but his informal schooling started the first time his plane was hit by enemy fire. Anything that affects the normal flying qualities of an aircraft instantly makes the rest of its flight an experiment. Is the plane on fire? Is it controllable? Can you make it back to your base? Should you eject, risking death or capture?

These are issues test pilots must address and solve instantly. Combat pilots face the same split-second, life-and-death decision making, with the added factor of other people trying to kill them.

Glenn's zeal to get close and personal with his targets allowed Korean gunners to turn him into a test pilot more than once.

He had not been in Korea long when, on a bombing run with three other jets, Glenn spotted an anti-aircraft gun firing at them from the right. Although Glenn was not the flight leader, he announced he was going after it and rolled into the attack. Glenn went low, all but scorching the treetops as he roared across the valley at 400 knots. He kept his jet aimed at the gun, firing away, pulling out just in time to avoid following his shells right into the target. At that moment, another gun across the valley opened up and ripped a hole the size of a dinner plate in the Panther's tail. The shot wiped out its elevator trim control and the Panther started nosing toward the ground. "It was the closest I ever came to crashing in my life," Glenn wrote in *We Seven*.[6] Instantly — reflexively — Glenn yanked back the stick, just in time to avoid planting the straight-winged jet in a rice paddy. The elevator responded poorly, but Glenn got the plane to climb. At 10,000 feet, he felt out the controls and found the Panther still controllable. He flew it safely back to base.

On another mission, one of Glenn's own bombs saved his life. The U.S. was using napalm again, and Glenn's unit was hauling heavy bombs full of the fire jelly under its wings. "We did a lot of the first, real heavy-duty napalm stuff off of jets in Korea," Glenn recalled.

On one mission, Glenn was just rolling into a bomb run when he felt his jet lurch.

Smoke billowed from the right wing, and Glenn had been hit. Glenn immediately turned toward the sea, figuring it was better to ditch near friendly Navy forces than eject over enemy-occupied land. The shell had blown away a three-foot section of his wing and damaged his trim control, but Glenn later learned he was lucky he still had a wing at all: the napalm tank under his wing had taken the hit, absorbing much of the blast and protecting Glenn's plane. Glenn coaxed the jet up to 15,000 feet and began testing it to see how much control he had; he even slowed the jet down to find out whether he would still be able to control it when he landed. Once again he made it safely home, where the ground crew found nuggets of shrapnel in the jet's fuel tank, but the self-sealing tank had kept enough fuel for Glenn to make it back. He kept some pieces of the shrapnel for souvenirs.

Glenn flew sixty-three missions with the 311th. Still, he had yet to tangle with another airplane. Glenn pressed to make his own opportunity. It finally came — from the Air Force, of all places. The Navy pilot turned Marine found a chance to join the Air Force's 25th Fighter Squadron as an exchange pilot. It was also a chance to fly the elegant, new F-86 Sabre, the swept-wing fighter that was going up against the equally advanced, Soviet designed MiG-15.

Although Glenn was now flying a jet designed for air-to-air combat, he wasn't finished with ground targets. Returning from a mission to escort bombers, Glenn followed his squadron commander, Col. John Giraudo, in a strafing run on a truck convoy. The valley erupted in anti-aircraft fire. Giraudo's plane was hit. He tried to make for the sea, but his crippled Sabre jet was out of control. He punched out over enemy territory. Glenn crisscrossed the area, trying to locate Giraudo and keep enemy troops away, but he was running low on fuel. He calculated how much fuel he would need to make it close enough to friendly territory to glide to a base or bail out. At the last moment, he climbed to 40,000 feet and aimed towards home. The engine soon flamed out, and Glenn's Sabre jet became a glider. Luckily, a tailwind extended his range just enough to glide back to his home base.

That wasn't the end of the adventure for Glenn. He had radioed ahead, telling the base to have a jet fueled and running for him to lead a rescue party back to Giraudo. "The boys in the squadron tell me John came in, rolled to the end of the strip, jumped into the other ship and took off like a man going to put out a fire," Giraudo said later.[7]

It was too late — Giraudo was captured almost immediately — but it illustrated Glenn's loyalty and tenacity. The story had a happy ending: Giraudo was released in a prisoner exchange shortly after the cease-fire on July 27, 1953, and Glenn was there to greet him.

Now, whether the "MiG Mad Marine" title was a joke as Glenn now claims or not, he earned the nickname on his first encounter with enemy jets. Leading a flight at 23,000 feet near the Yalu River, Glenn spotted MiGs far below at 2,000 feet. Glenn and his wingman swept into a steep dive, closing from behind at high speed. Glenn poured a long burst into his quarry until it erupted in flames and plunged to the ground. Glenn returned to his base, the Sabre jet's nose blackened with smoke from its six machine guns. Instead of counting holes, the ground crew painted a red

US NAVY

After WWII ended, Glenn returned to the United States before he was re-assigned to Guam and China, where he flew the F4U-1.

star on the side of the jet.

His second kill came just a week later while Glenn was leading a flight of four Sabre jets. He spotted four MiGs crossing the Yalu for North Korea, and his flight went on the attack. Just as they were closing on the MiGs, more MiGs appeared. Glenn's four-ship patrol suddenly found itself facing sixteen MiGs. Another four Sabres joined the fray, and suddenly the sky was dotted with twisting, turning fighters.

Glenn and another MiG took turns switching roles as pursuer and pursued. Glenn maneuvered for a shot, but before he could fire, his wingman turned on the MiG and blasted it.

But the wingman's engine was losing power. He turned for home, limping along at low altitude, while Glenn flew above to protect him. Six MiGs spotted them and dived in single file like a line of hungry raptors. Glenn turned on the lead plane and fired a long burst. The enemy leader broke off, and the others followed. Glenn rolled in behind the last one and put a volley of bullets into its fuselage. The MiG fell away, trailing smoke, and crashed into a hillside.

Time after time the MiGs started for them. Each time Glenn turned them away with a fusillade of bullets. Glenn's wingman finally radioed that he was safely close to base. Glenn opened fire on another oncoming MiG and ran his guns out of bullets. Alone and out of ammo, he turned into a steep dive and raced for home, making a good case for the fighter pilot's adage, "Speed is life." The Sabre jet could outrun the MiGs, and the enemy jets quickly gave up the chase.

Glenn scored his third air-to-air kill on July 22, when his flight of four quickly shot down three MiGs.

As it happened, they were the last three MiGs shot down before North and South Korea agreed to a cease fire on July 27.

Glenn had prevailed in his last gunfight, but the tenacity of the "MiG Mad Marine" was just beginning to show.

NOTES

1. Clair Stebbins, "First Kill Thrills Glenn," *Columbus Dispatch*, no date.

2. Frank Van Riper, *Glenn — The Astronaut Who Would Be President* (New York: Empire Books, 1983), p. 85.

3. Pierce and Schuon, *John H. Glenn—Astronaut*, pp. 29-31.

4. Van Riper, *Glenn — The Astronaut Who Would Be President*, pp. 100-102.

5. Pierce and Schuon, *John H. Glenn—Astronaut*, p. 33.

6. Mercury Seven astronauts, *We Seven* (New York: Simon & Schuster, 1962), p. 38.

7. Pierce and Schuon, *John H. Glenn—Astronaut*, p. 38.

GLENN, USMC

Glenn's destiny seemed at times to be guided by fate. One of two pilots involved in Project Bullet, Glenn was the only one to finish the cross-country race against time because the other pilot's plane developed engine problems.

CHAPTER 4
THE TEST PILOT

At times you can almost believe John Glenn knew spaceflight was coming before it came. It's as though he had magically foreseen Project Mercury years before it dawned in 1958, just after NASA itself was born. He was so driven, so accomplished, and so unerringly accurate in charting his life toward the unseen point in science where pilots would become astronauts that looking back on it all gives an illusion of Glenn as clairvoyant. From Army enlistee to Navy to Marines, Glenn skipped quickly into position to fly the hottest planes on Earth. Aching to be an attack pilot, he had settled temporarily for a gig learning multi-engines early in his military career in case he needed to make a living as a civilian cargo pilot. But then he and his buddy Tom Miller talked their way into slots in an attack squadron anyway, and Glenn ended up with an unusually broad range of flying experience that in early 1954 proved to be a real asset.[1]

Diplomacy had sketched its uneasy peace along the 38th Parallel in the summer of 1953, sending Maj. Glenn home just months after he was introduced to the thrill of air-to-air combat. Liberated from the yeoman's work of protecting bombers or dropping napalm in ground attack runs, as Marine fighters so often did, Glenn shined brightly as a loaner pilot with the Air Force in their sexy F-86 Sabre jets. His machine guns chewed into many a Chinese MiG, and Glenn left combat flying with a chest full of medals, hours of cockpit time in jets, and an appetite whetted for still more exhilarating duty.

He found it in the Navy's test pilot school.

The U.S. Naval Test Pilot School occupies a scenic corner of the Patuxent River Naval Air Station, bordered by the river on one side and the Chesapeake Bay on the other. The air strip itself launches planes out over the tidal marshes that buffet Maryland's Atlantic coast, just as it has since the school opened in 1948. It is here that top fliers master the techniques and delve into the art of taking a plane into the sky and finding its physical limitations. Today, students spend a year in the school and much of that time is in the classroom. In Glenn's day, the program was half as long, but they did just as many test flights, roughly sixty-two.

Glenn entered the program in January 1954 at a disadvantage. He had never completed his chemical engineering degree at Muskingum College because of WWII, which meant he hadn't taken some of the higher level academics that were essential to an understanding of advanced aerodynamics. At Patuxent in the 1950s, pilots were neck deep in advanced aerodynamics seven days a week for those six long months. The curriculum included nearly 100 hours of classroom work

NASA

Each step in Glenn's life, from Marine flyer to astronaut, seemed to lead perfectly to the next one.

in algebra, calculus, and physics, in addition to another 200 hours on aircraft engines and aerodynamics.[2]

"He had a hard time because he had not concentrated on math in college," said Tom Miller, "and they were fooling around with analytics and geometry and trigonometry and stuff that made it pretty damn difficult for him."

Using slide rules and sharp pencils, each student crunched a zillion numbers and plotted them on graphs to create precise, numerical performance evaluations of the planes they flew. Fuel consumption at assorted speeds and altitudes, rates of climb, rates of turn, agility. Everything a data point. Later, computer advances would allow high volumes of such information to be recorded instantly, automatically. Instructors and other pilots would eventually be able to read the data from planes still in the air. But back in 1954, the most important recording device in a test pilot's cockpit was the notepad strapped to his knee. The thirty-member classes were broken down into work groups of about six each. They met in a generic classroom to hear lectures and to prepare their flight reports. The rest of the time they flew. Or studied.

Perhaps a little anxious about the course work that lay ahead of him, Glenn arrived at Patuxent without his wife and children. They stayed behind with her folks in Ohio, and Glenn lived in the base's Bachelor Officer's Quarters, which freed him up to put in long, long hours.

"He moved into the BOQ on the base, but he would study every night in the study room," remembers retired Vice Adm. James Stockdale, a classmate in the pilot school. "And it soon became common place, if you forgot what the lesson was for the next day — and it was a pretty rigorous school — the phone would ring [in the study room] and John would answer, and you'd say, 'What is it we're going to do tomorrow, and what's the pages we've got to read?'

"And he'd always know it."

Glenn was not the only achiever in Class 12 of the Naval Test Pilot School at Pax River, as they call it. This was a collection of the Navy's best flyboys, so you could expect that from their ranks would emerge a few leaders. Stockdale is one of them, too. He achieved the rank of vice admiral having won the Congressional Medal of Honor for his almost unbelievable bravery in leading American defiance against torture during his seven years as a prisoner of war in Vietnam. After his release in 1973, Stockdale commanded the Antisubmarine Warfare Wing of the U.S. Pacific Fleet. Later he moved into more academic roles, including tenures as president of the Naval War College and then the Citadel. In 1992, he ran as the vice presidential running mate of Ross Perot.

But back in 1954, Stockdale was a young attack plane pilot and a graduate of the Naval Academy at Annapolis. Although he had logged plenty of hours flying on and off carriers, Stockdale, oddly enough, had never flown a jet. He did, however, have the math and physics background Glenn lacked, so together they arranged a swap of skills. Stockdale would tutor Glenn in the books, and Glenn would introduce Stockdale to jet flying. Glenn delivered his half of the deal on an overcast evening in Maryland, early in the school term.

It was a windy, almost mysterious night. Perfect for Halloween, but this was late winter. Any old time now, it could rain. Out of nowhere, Glenn popped in on Stockdale. "We're going flying tonight. We're going down to Miami," Glenn said.[3]

Now, Miami is a few hours away from Pax River, even in a jet. But Glenn didn't take no for an answer, saying they both could use the flight time and experience. "There's an air station down there so we can refuel for the flight back," Glenn said. "Just up and back, c'mon." Before you know it, they were off, ripping down the East Coast with night deepening around them and thunder clouds trying

hard to break open with a squall. The plane was a two-seater, a trainer, with Stockdale in the backseat and Glenn up front at the controls. Not much happened on the flight down, maybe a little small talk. They landed in Miami, refueled, and filled out the yellow sheet, where mechanical problems would have to be noted if there had been any. The weather was no better down south, breezy and ominous. And on top of that, the Miami mosquitoes are fierce. Glenn and Stockdale got the plane turned around quickly.

Rain started as they walked back to the plane, and the drops glinted in the airfield lights like schools of fleeing bait fish.

"You know," Glenn said coolly, "you're going to fly back."

"Oh yeah?" Stockdale replied.

"Yeah, you can handle it, I know."

Without further discussion, Glenn hopped up into the rear seat, and Stockdale took the pilot seat. Once Stockdale was in, Glenn stood up behind him, clicked on a flashlight, and leaned over Stockdale's head to give the basics. "OK, now we're going to start this thing. Now you see that button over there?" The flashlight spots it. "OK, push that, and when the RPM comes up to twelve percent you take that throttle and go around the horn and we'll get it started. . . ." And Stockdale was sitting there, canopy open, rain misting around him, Glenn talking in his ear, and a flashlight dancing across his control panel.

Minutes before take-off. In a plane he had never flown before. Stockdale was keenly aware of how strange this was but said nothing and rolled the plane out into position on the runway.

Moments later they fired into the ink sky over the Atlantic Ocean. Stockdale felt the plane, with Glenn studying him over his shoulder. Counting off the minutes, Stockdale banked north for Pax River.

"You can't imagine what a boost this gave

my morale and my confidence," Stockdale said.

They ended the night at 2:30 a.m. back at Stockdale's house over bacon and eggs. Stockdale slipped down the hall to tell his wife he was home, and to tell her he'd brought this guy from the flight school back with him. (There's no place to scrounge up a little dinner at 2:30 a.m. in the BOQ.) When the Stockdales returned to the kitchen, Glenn was in Mrs. Stockdale's apron and had the bacon popping in a fry pan.

"That was really the basis of a very enduring, long-range friendship," Stockdale said.

Nothing really is the equivalent of combat flying for the way adrenaline washes in during those life and death moments. You can't replicate the combination of shooting and being shot at in the 3-D world of aerial combat, and you can't simulate the exhausting way your brain and hands and heart respond in a kind of explosion of focus and precision. You probably wouldn't want to. But the test pilot school did a fine job of wearing down these young pilots, nearly all of them combat veterans. War-time flying has horrifying moments, but they're punctuated by days, sometimes weeks, of boredom. In test pilot school, the action never ceases.

"I'd say you don't have that discrepancy between the average and the high, but the level of average [activity] there is high in test work and it goes on and on, day after day," said Frank Posch, who also was a test pilot classmate of Glenn's.

It's not combat but, "It's approaching that. It's approaching the high tension, high awareness, high energy, high concentration," Posch said.

Later, Glenn would tell the story of how he once lost a chunk of his right wing while flying low-altitude combat maneuvers at the test school. He was really putting the fighter through its paces, and the resulting wing fracture cost him a large part of the surface area needed to give the plane lift. As he'd done in real combat after being damaged by artillery, Glenn calmly ran his speed up and down until he had a good feel for the limits of his control. No problem landing, he decided, I'll just have to come in at a little higher rate of speed.[4]

Another part of the burden was the need to be skilled at flying all kinds of machines, Posch remembered. "It was very difficult for some of these guys. The multi-engine people suddenly found themselves flying a fighter airplane and the fighter pilots found themselves flying a four-engine thing. . . . And that was part of the general screening process there. You know, every once in a while somebody would drop out, just too much, you know, couldn't handle all the confusion."

But not Glenn. The Marines had trained him on multi-engine planes shortly after flight school, then let him switch over to an attack squadron. Instructors at the test pilot school counted on the strain of switching plane types as a means to weed-out weak pilots, but Glenn had mastered that skill before he even got there.

In addition to the grind of daily reports on planes, each pilot had to write a term paper by the end of class. Typically, these were not stellar examples of scholarship. Posch bluntly called them a collection of "forgettable" papers that couldn't be much better than that because the day-to-day work load was so heavy nobody had time to put much thought into them.

"To tell you the truth, I can't remember exactly what the title of mine was right now," Posch said. "I could dig it out if I had to."

But Posch remembers Glenn's.

"It was the one that showed some real imagination."

US NAVY

Several times, Glenn submitted a proposal to his superior officers that would allow him to attempt a national air speed record across the United States. Glenn finally received approval and was part of Project Bullet that flew two F8U jets from coast to coast.

In his paper, Glenn pointed out that weather patterns, high and low pressure systems, needed to be taken into account when calculating flight data because of the way they affect altitude readings, Posch said. Since altimeters measure air pressure rather than distance from the ground, flying into a low pressure weather system feels to an airplane as though it's diving, which influences the consumption of fuel and other things.

"Then he calculated through on this and found out that this could make a significant, not an overwhelming, but it could make a significant difference in the accuracy of your data," Posch said.

Posch said Glenn's paper fostered a change in thinking among pilots and instructors.

"Baseball teams and naval aviators aboard carriers: they sort themselves out very quickly," Posch said. "There's always a small group of exceptional ones, and John was always part of that exceptional group. Just didn't have to be with him more than a week or two and you sensed that."

Of course, fast airplanes, even jets, did not represent the sharpest cutting edge of air travel. A conglomerate of America's civilian and military scientists were refining the practical applications of the still awe-inspiring rocket. Many of those scientists had been appropriated from Germany, which was a leader in aviation even as it was losing World War II. Before Glenn returned from Korea, America's pursuit of battlefield rockets had escalated among priorities at the Department of Defense. The Pentagon's budget for rocket research crossed $1 million for the first time in 1951, then passed $1 billion just six years later.[5] Scientists and engineers were chasing the dual goals of making warheads smaller while making rockets more powerful, hoping a perfect combination of the two would lead to an intercontinental ballistic missile. But along the way, the balance was lost, and America emerged as masters of miniaturized warheads, while losing ground to the Soviet Union in the competition to build bigger rockets.[6]

And although the military uses of rockets were obvious, at this same time there was an interest in the scientific community for using a rocket to punch a satellite into Earth orbit for the first time. The civilian National Advisory Committee for Aeronautics, which had been founded and run by the likes of Orville Wright and soon would be replaced by NASA, was moving further and further into rocket research with an eye on the highest heavens. The NACA even roughed out a plan to put a man in space using the best-understood principals of aerodynamic science at the time. Namely, a cone-type spacecraft on the end of a ballistic missile, designed with one blunt end for reentry into Earth's atmosphere.[7]

Glenn had moved on from Pax River to a job as project officer in the Navy's Bureau of Aeronautics in Washington, D.C. The Marine Corps is a branch within the Department of the Navy, so Glenn found himself working in the overarching management structure of both the Navy and Marines. It was a bit of a desk job, but the young major was assigned to the Fighter Design Branch, where he was able to engage his engineer's mind and combat pilot's heart by working on the development of war planes.

Figuring they'd stay put for a few years,

the Glenn family went looking for a suitable place to live and settled on Arlington, Va., a hilly green suburb of the nation's capital that can look like New Concord. As had happened so often in Glenn's career, his buddy Tom Miller landed a job in Washington at the same time, so they went house hunting together. Ultimately, the two Marine majors concluded that they couldn't afford to buy a house in the neighborhood of their choice — near a highly regarded junior high school — but got a deal on a pair of adjacent lots directly across the street from the school. The matching ranch-style houses were built into the side of a hill so there was no basement but rather a gloriously wood-shaded downstairs living area. The houses, and more precisely their vast front yards dominated by a pair of enormous oak trees, became the focal point of America's fascination with the space program just a few years later.

Meanwhile, Glenn was working as the Navy's point man on a faster-than-sound bullet of an airplane called the F8U Crusader. Shaped like a cigar, with wings that started far from the nose and swept back steeply, the Crusader was a fine example of the latest in airplane design that had improved exponentially in the decade since the sound barrier was broken. In the Crusader, the Navy and Marines at last were getting a fighter jet that could fly as fast and high as any plane in combat, something difficult to achieve because of the structural reinforcements required on planes flown off ships. While working on the plane's development and flying it over test ranges back at Pax River, Glenn had an idea. Now, these were the days when aviation was bursting with new ideas, and setting records was more than just macho whimsy, it was part of the practical side of pushing new technology. Just how much can this thing do? So Glenn's idea was to fly a stock Crusader from Los Angeles to New York, and do the entire trip at a speed faster than sound. Faster, actually, than a .45 caliber bullet. He figured it would be a good way to test the engine's endurance at wide open speeds, and if everything went well it could snare a Marine Corps pilot a nice speed record.

The Navy said no.

But Glenn was evolving into one persistent Marine. No was not an acceptable answer, so he reworked the flight plan and submitted the request again. No. So he did it again.[8]

'Operation Bullet' finally took off from the Los Alamitos Naval Air Station south of Los Angeles at 6 a.m. on July 16, 1957. Glenn and another pilot, a Navy Lieutenant, soared off in a pair of matching

Maj. John Glenn Jr,. took just 3 hours, 23 minutes, and 8.4 seconds in 1957 to fly from California's Los Alamitos Naval Air Station to New York's Floyd Bennett Field — nearly 22 minutes faster than anyone had previously made the 2,460-mile trip. Glenn averaged about 850 miles per hour — faster than the speed of sound.

Crusaders. (They used the reconnaissance-designed photo version of the fighter, which had cameras instead of guns but was otherwise identical to the combat jet.) But the Navy Lieutenant had trouble hooking up for the first of three scheduled mid-air refuelings and had to bail out, so Glenn quickly became the one and only shot at making a record. Flying nearly ten miles high at just over 1,000 miles an hour, Glenn aimed for New York. But then on the second half of the trip, Glenn too ran into trouble connecting with a refueling tanker, this one over Indianapolis. Refueling was a huge burden on the project, and perhaps the greatest obstacle to setting a record because

Minutes after he stepped from the plane in which he set the cross-country mark, Maj. John H. Glenn Jr. is surrounded by his family — Lyn, 10; David, 11; and Annie Glenn — at Floyd Bennett Field.

the Navy still had only propeller-driven tankers that lumbered through the sky at under 300 miles an hour. For each rendezvous, Glenn had to begin slowing down thirty-five miles away, losing precious minutes each time. But then over Indianapolis he got a false radio signal that sent him circling the wrong way to meet up with the tanker. When the error was discovered seconds later, Glenn nearly pulled the Crusader apart as he banked her back toward his fuel supply. When this last refueling was finished, Glenn throttled up in a hurry.[9]

"From somewhere along there in the Midwest on into the East Coast we left a supersonic sound boom dragging pretty much behind us all the way across the country," Glenn said during an autumn speech after the flight.

The sonic boom reportedly collapsed a ceiling in a house in Terre Haute, Indiana, and cracked windows in Pittsburgh. Glenn's family and friends in New Concord had scurried outside to catch him flying directly overhead, which Glenn did, thumping his pretty hometown with the shockwave of a jet doing Mach 1.35. As Glenn soared away, the phone rang at his parents' house. It was a horrified neighbor.

"Oh, Mrs. Glenn, Johnny dropped a bomb! Johnny dropped a bomb!"[10]

Glenn arrived in New York 3 hours, 23 minutes, and 8.4 seconds after leaving L.A. He set a transcontinental speed record as the first man to average supersonic speed all the way across America and earned his fourth Distinguished Flying Cross from the Navy. The Marines nominated him for the Harmon International Trophy, an aviator's prize that had previously been won by Chuck Yeager and Howard Hughes. In the nomination letter all praise went to Maj. Glenn.

"It might be worth mentioning that it was largely through repeated efforts on the part of the major that the Navy and the Defense Department permitted the flight to be made," the Harmon application reads. "Three times he had submitted detailed proposals and flight statistics before, on the fourth try, his efforts bore fruit."

Glenn didn't win the award, but a hallmark for tenacity was cemented.

And besides, he was moving on to other things. A talent scout for CBS had spotted him while he was shopping with Annie in New York and asked him if he'd ever thought about doing television.

"Some gal just walked up to him. . . ." remembered Tom Miller. "She didn't recognize him or anything. She just said, 'How'd you like to be on a show?'"

Glenn did a couple of game shows then, but was most successful when he teamed with youngster Eddie Hodges to win about $25,000 on the television game show "Name that Tune" in October 1957. Hodges went on to a Broadway career that included a role in the musical "The Music Man." Glenn, the former Glee Club member and a life-long devotee to a good barbershop quartet, did his part by quickly recognizing such tunes as "Far Away Places."

But TV wasn't really the point. The Marines may have liked the P.R. — they posted his appearances on the headquarters bulletin board — but news was brewing. Cold War news. Space news. In the same week that Glenn was grinning it up on "Name That Tune," the Soviet Union launched into orbit the very first man-made satellite. On October 4, 1957, Sputnik began circling the world. The significance of this event almost can't be overstated. It wasn't simply momentous for the world, or embarrassing for America. It ran deeper than that. To the United States and her allies, the launch of Sputnik was chilling. The *New York Times* needed three full sentences stretched the width of its front page to produce a headline that could capture what had happened:

SOVIET FIRES EARTH SATELLITE
INTO SPACE;
IT IS CIRCLING THE GLOBE
AT 18,000 M.P.H.
SPHERE TRACKED IN 4 CROSSINGS
OVER U.S.[11]

No doubt, American scientists were surprised by the launch. Two years earlier, the U.S. had proudly announced its plan to launch a science satellite sometime during the celebration of the International Geophysical Year, which ran eighteen months beginning in the summer of 1957. Scheduling it for the IGY was supposed to emphasize the civilian, rather than military, nature of America's interest in space. And while U.S. engineers were optimistic that they would meet that timetable — certainly they could make a space shot before the end of 1958 — they didn't expect the Soviets to leap so quickly ahead.

But what really impressed the Americans, downright frightened some of them, was the size of the Soviet's beeping metal ball. The Russians had managed to put a 183-pound satellite into space, while America was still fine-tuning its plans to launch something weighing three pounds. The outside range for America's Vanguard rocket program of the day, the very best they hoped to achieve after a series of progressively bigger launches, was still only twenty-two pounds. At the International Astronautical Congress, which was meeting in Barcelona at the time of the Sputnik launch, a U.S. military official was quoted in the magazine *Aviation Week* as saying, "If it weighs eighteen pounds they're ahead of us; if it weighs 180 pounds, I'm scared!"[12] It had only been two years since the Soviet Union detonated a useable H-bomb (something the U.S. did a year later), and now

it was clear they were strides ahead in developing a rocket with the horsepower to launch such a warhead across oceans. No matter what else could be said about the meaning of Sputnik I, the very minimum was this: the Soviet Union had built one monster rocket.

And it didn't end there. President Eisenhower remained his stately, grandfatherly self amid the public furor, proclaiming the world's first space shot was less ominous than it seemed. But a month later, the Soviets did it again. Sputnik II weighed more than 1,000 pounds and carried a living, breathing dog, which survived temporarily in orbit before the satellite overheated when the booster failed to separate.[13] By December, the U.S. felt compelled to act and did. The Vangard 6, carrying a three-pound sphere which Soviet Premiere Nikita Khrushchev referred to as an "orange," climbed from its launchpad in Cape Canaveral on December 6. Unfortunately, it stopped climbing at about three feet and exploded. Thus was born, "Kaputnik."

But while the world slept under a Soviet Moon, as the phrase of the day went, and everyone fretted about a growing "missile gap" with the Soviets, scientists at the Langley Research Center in Virginia inched forward with their program for putting a man into orbit. Engineers there had called the Bureau of Aeronautics and asked if they could borrow a pilot, essentially to take a seat in some of the earliest space capsule designs and see how it all looked.

"They had the engineers, they needed somebody with more flight experience to come down and work with them on hand control or some things like that," Glenn said. "And they had asked the Bureau of Aeronautics to assign someone to come down and be a bit of a guinea pig to help them work out some of these things."

Glenn volunteered.

"It was just interesting," he said. "It was

a natural follow on to what I'd been doing for the previous four years, or three and a half years in test work. We'd been testing airplanes, and I'd been doing test work on some of the first of our supersonic airplanes."

The assignment sent Glenn south to Langley, and also north to Johnsville, Pennsylvania, where the Navy had a large and powerful centrifuge. By whirling a man in the centrifuge, scientists could simulate the accelerating power of a rocket launch or reentry thrust. Langley researchers had been focusing on the issue of how high the G-load would be on an astronaut burning back into the Earth's atmosphere after an orbital trip, and they wanted some data on heavy-G runs. Glenn did the runs, provided Langley with the data they wanted, and kept a copy of the information for himself. It wouldn't be long before Glenn would put that file to use.

Soon after that, those same researchers sent Glenn to St. Louis, where aircraft maker McDonnell Corporation was building the first generation of the Mercury capsule. Before it was known to the public as the Mercury capsule. Glenn represented the Navy in discussions of the mock-up design.

Once again, Glenn had found himself in the right place at the right time, doing the right thing. Of course, this had happened because he'd volunteered again, too. In the book *We Seven*, which Glenn and the others wrote back in 1962, Glenn said that while he did this work for the space researchers at Langley, he had "no inkling" that he might become an astronaut himself.

It just looks like he did.

NOTES

1. Interview with Ret. Lt. Gen. Tom Miller, in Arlington, Va., June 13, 1998.

2. United States Naval Test Pilot School, Historical narrative and class data — 1945 to 1983, pp. 171-172.

3. Interview with Ret. Adm. James Stockdale, by telephone, June 1998.

4. *We Seven*, p. 41.

5. *Orders of Magnitude, A History of the NACA and NASA, 1915-1990*, The NASA History Series, p. 43.

6. Ibid., pp. 44-45.

7. Ibid., p. 46.

8. USMC essay supporting the nomination of Glenn for the Harmon International Trophy, Archive of the USMC Historical Center, Navy Yard, Washington, D.C.

9. Transcript of speech by Maj. John Glenn to the National Exchange Club Convention in Atlantic City, NJ, September 7, 1957.

10. Ibid.

11. Abstract of James J. Harford, "Korolev's Triple Play: Sputniks 1, 2, and 3," adapted from Harford's *Korolev: How One Man Masterminded the Soviet Drive to Beat America to the Moon* (John Wiley: New York, 1997), p. 7.

12. *Sputnik and the Birth of NASA: An Overview*, p. 2.

13. Abstract of "Korolev's Triple Play," p. 8.

The Mercury astronauts. Front row (l. to r.) Walter Schirra Jr; Deke Slayton, John Glenn Jr., and Scott Carpenter. Back row (l. to r.) Alan Shepard, Gus Grissom, and Gordon Cooper.

CHAPTER 5

THE ASTRONAUT

Originally, the first astronauts were going to be chosen from the expanse of civilian America. The scientists at NASA were looking for daredevils and brave hearts, and they didn't care where they came from. A leader of the search team said that cliff climbers, scuba divers, recreational parachutists, and "people who went in for stressful sports" would be encouraged to apply. Near the end of 1958, with NASA just a couple of months old, the selection group drafted an announcement seeking applicants for the civil service position of research astronaut. The notice included a job description, an annual pay scale of between $8,330 and $12,770 depending on experience, and a requirement that some reputable organization sponsor every candidate.[1] NASA wanted to avoid independent thrill seekers who lived too far out on the edge. Bravery was critical, but they didn't want courage teetering on the cusp of crazy. The applicants would have to show they had a record of accepting hazardous duty willingly, of tolerating severe environmental conditions, and of reacting well in stressful situations. Part of the announcement NASA planned to distribute said: "These three characteristics may have been demonstrated in connection with certain professional occupations such as test pilot, crew member of

Glenn was the oldest of the seven astronauts.

experimental submarine, or arctic or antarctic explorer."

But in late December, all that was scrapped. The announcement never left NASA's offices. President Eisenhower, who had insisted the space program be a civilian enterprise, abruptly decreed over the

The United States named their first astronauts on April 9, 1959 — Malcolm "Scott" Carpenter, Leroy "Gordon" Cooper, John H. Glenn Jr, Virgil "Gus" Grissom, Walter Schirra Jr., Alan Shepard, and Donald "Deke" Slayton.

Christmas holiday that only military pilots would be considered for the civilian space agency's first big program, Project Mercury.[2]

"It happened that President Eisenhower didn't like the idea of having a nationwide screening process for astronauts," said Charles Donlan, who was deputy director of Project Mercury and led the search for astronauts. "He thought the program was probably going to go away after the excitement of Sputnik so he vetoed that wide-scale attempt to screen astronauts and said, 'Take them from the files of the Defense Department. Take them from the graduates of the flying schools. . . .'"

Well, this certainly made things easier.

NASA put together a list of minimum requirements from height to weight to flying hours, and they whittled the military pilot "files" from 508 to 110.

John Glenn's name was among them, but on paper he had a couple of problems. He still lacked a bachelor's degree — a minimum requirement — and at age thirty-seven was only three years from the top of NASA's age bracket. He also may have been a little too tall. Today Glenn dismisses this story as silly, but his buddy Tom Miller swears that as NASA was weeding through its applications, pulling out everyone taller than 5' 11", Glenn was stacking books on his head to compress his spine.

"Well, he learned to scooch down. On his medical record, I'm almost positive of this, I can't remember seeing the record but I think he was listed as six foot," Miller said. "In fact, he used to mark a place on one of the wall facings . . . to see how far he could get below it. If you ask him about that today, he'd probably deny it."

With his background in test flying, his exemplary record in combat and his lengthy experience with jets, Glenn overcame his academic deficiency and made the paper cut. He was one of only five Marines among that 110.

And then fate gave him a nudge. To make the process easier to manage, Donlan randomly divided the 110 names into three groups and invited groups one and two — roughly seventy pilots — to Washington for interviews in early 1959. These individual meetings with the pilots not only gave Donlan and his crew a chance to begin sizing everybody up, but it also did the same for the pilots. Only basic information was known about the call for astronauts, so NASA's people needed to outline the nature of Project Mercury in case any of the qualified pilots didn't want to volunteer. And some didn't. Heading off to join a fledgling space program for the chance to follow a monkey into orbit didn't look like a brilliant career move to everybody, especially among a group of young officers with bright military futures in front of them. And it was around this time that a group of experimental aircraft pilots began mocking the job astronauts could really do in a floating space capsule by saying they'd be nothing more than "spam in a can." So by March, the group of seventy pilots had dropped to thirty-six, either because they quit or were eliminated. (Six had actually grown too tall.)[3] Still, Donlan thought this gave him a large enough pool to

select a final set of astronauts, so he never even interviewed the last group of pilots.

Fate's push, or maybe dumb luck, was all that had put Glenn in with the pilots that met Donlan instead of in the group that never did. Glenn's Presbyterian faith fortified him throughout his life for moments like this by investing him with the belief that if he just did his part, then fate would do the rest. He called it the "fifty-fifty proposition," and here again, he had done his part and so had fate. Which meant it was his turn once more to deliver. Now, becoming an astronaut was simply a matter of being better, of being exceptional among those delivered by fate or luck or a "higher power," as Glenn called it, to Donlan's office for consideration.

And so he did. He began making an impression instantly. The pilots had been told that Project Mercury was a civilian program, so they should just relax and wear street clothes for the preliminary interview. But that was too pedestrian for Glenn.

"I remember John Glenn first came into my office for the initial interview in his Marine uniform," Donlan remembered. "And he had a brown envelope under his arm."

An envelope? Nobody else brought a folder. Donlan asked what he was carrying.

Centrifuge runs, Glenn told him.

In the brown envelope, Glenn carried the data on high-G centrifuge runs that he had amassed for the Langley researchers when they sent him to Johnsville just a year or so earlier. Glenn explained that he knew NASA was worried about the heavy strain of reentry, and he just wanted the astronaut selection people to know he had been involved in early research on the topic.

"He was the only one who had material like that," Donlan said. "In fact, Glenn was the only one, after seeing the drawings that I had of the Mercury [capsule], who asked if he could come back that evening and pore over them.

"Those are the kinds of things that we looked for. The dedication to the program."

The selection course that followed for Glenn and the other astronaut candidates in the group of thirty-six was rigorous almost to the point of comedy. Scientists had utterly no idea what space would do to the human body, so they devised a battery of physical and psychological tests that stretched the imagination. At Wright-Patterson Air Force Base in Dayton, Ohio, and a laboratory in Albuquerque, New Mexico, the candidates were pummeled with tests. Wright-Patterson and its doctors played a large role because this was where the Air Force housed its space program, called Man In Space Soonest (MISS), which had been making progress on the topic before being closed down to put full military support behind Mercury. During the exams, ice water was poured into their ears to disrupt balance and they were locked in silent, lightless chambers for hours on end.

"We went through heat chambers up there where they heated your body up until your core body temperature got up to a point where they thought it was dangerous to go beyond. . . ." Glenn once told a space and medicine symposium in Washington, D.C.

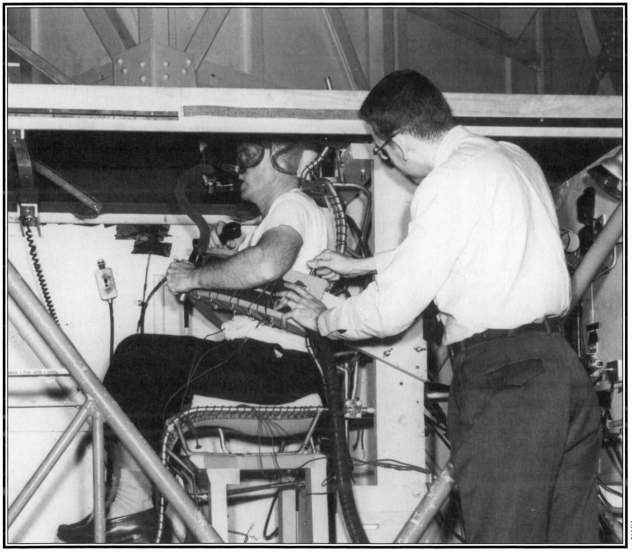

Glenn knew the importance of centrifuge training and brought his centrifuge test results to his first astronaut interview.

"We went through sound chambers where the frequency and amplitude were varied to the point where it'd make your whole body shake, literally."

Donlan said, "I viewed it as these medical people finally had a batch of top-notch guys, and they kind of used it as a research program of theirs to see how normal people behaved in some of these odd tests they gave them."

Glenn viewed it as a hurdle. Using the tenacity he had displayed so well in getting the Navy to let him fly for a transcontinental speed record in 1957, Glenn marched through the physical and psychological examinations.

The astronaut candidates were asked to complete sentences such as "I am sorry that . . . ," and "I can never . . ." And respond "True" or "False" to such statements as, "Sometimes I feel like cursing," and "Strangers keep trying to hurt me."[4]

They looked at dozens of the Rorschach ink blot tests in which subjects explain what they see in a random splatter.

"They all looked like butterflies to me," Glenn said. "I kept telling them that whether it was right or wrong."

The exams were so thorough and so imaginative that when Dr. Randolph Lovelace — who as chairman of the NASA life sciences committee had led the team of doctors — was introduced at the press conference naming the Mercury Seven, his first words to reporters were, "I just hope they never give me a physical examination."[5]

That day was April 9, 1959. The group of

In addition to physical training, Glenn and the other Mercury astronauts went through imaginative, but thorough, physical examinations.

thirty-six had been whittled to seven, which actually was a bit of a compromise. The early plan was to choose twelve but Donlan thought that was too many. They weren't planning twelve flights, so why train a bunch of extras who would never get in the game?

So they picked seven.

Malcolm "Scott" Carpenter, 33; Leroy "Gordon" Cooper, 32; John H. Glenn Jr., 37; Virgil "Gus" Grissom, 33; Walter Schirra Jr.,

35; Alan Shepard, 35; and Donald "Deke" Slayton, 35, were introduced at a Washington press conference. Again trying to emphasize that this was a civilian program, their military titles — three captains, two commanders, one lieutenant, and Glenn ranked highest as a lieutenant colonel — were omitted from the introduction.

The first question was about their wives. All seven were family men, and they were asked if their wives supported their decision to embark on this dangerous new mission. It's a question Glenn would be asked again over the years, but never more often and perhaps more seriously than thirty-nine years later in 1998, when the elderly senator announced that he would return to space aboard a shuttle to help NASA study the effects of weightlessness on an aged body. The answer now is not much different from the answer then.

In 1959, Glenn answered that question like this: "I don't think any of us could really go on with something like this if we didn't have pretty good backing at home, really. My wife's attitude toward this has been the same as it has been all along through my flying. If it is what I want to do, then she is behind it, and the kids are, too, 100 percent."

In 1998, NASA Administrator Dan Goldin answered the question for Glenn, saying he called Annie Glenn before approving her husband's return to space. "You know what she said to me? 'This is what John wants to do. I've been with him for fifty-five years, I'm going to support him,'" Goldin said.

But in 1959, other questions required attention. A reporter at the press conference noticed that three of the Mercury Seven were smoking cigarettes and said, "I wonder what they are going to do for a cigarette when they get up there?"

Dr. Lovelace didn't have a firm answer because so much of the Mercury Program still was on paper. So much still was unknown. He said something about letting each pilot decide for himself. It would be some months before NASA engineers would decide to compensate for the ultra high altitude of a capsule in space by filling it with 100 percent oxygen rather than air. Of course, once that was decided, it was clear there most certainly wouldn't be any smoking in the capsules.

It was during this introduction to America that gung-ho Glenn began standing out. He was the redhead, sure, and the only one with freckles, which crossed his face and climbed his forehead until they dissolved into a receding orange crewcut. But beyond that, he was the most talkative, lacing his comments with anecdotes and personal philosophy. His answers were longest, and he went for the laughs where he could. When a question came about the most difficult test they took during the physical examination process, Glenn's delicate, understated response was greeted with howls and a simple chorus of "Amens" from the other astronauts.

"It is rather difficult to pick one," Glenn said, "because if you figure how many openings there are on a human body, and how far you can go into any one of them [the laughter began] you answer which one would be the toughest for you."

Said Cooper: "I think he answered it very well."

Schirra: "I think that goes for all of us."

What emerged of Glenn in this press conference was a sense that the man believed he was there among the Mercury Seven for God and Country, no joke. Yes, he was excited personally, but when questions of motivation and spiritual grounding and desire to be in the program crossed the room, Glenn talked as if all this was the culmination of a life-long mission. One reporter asked for a show of hands of those who were confident they would make it back from space safely, and as you would expect, all seven raised their hands. But Glenn raised both of his like a man surrendering, and to hear him talk at the press conference that day you might think he felt that way, too. Surrendering, to duty or history.

Glenn said, "In answer to this same question a few days ago from someone else I jokingly, of course, said that I got on this project because it probably would be the nearest to Heaven I will ever get, and I wanted to make the most of it.

"But my feelings are that this whole project with regard to space sort of stands with us now, as if you want to look at it one way, like the Wright Brothers stood at Kitty Hawk about fifty years ago, with Orville and Wilbur pitching a coin to see who was going to shove the

The astronauts had to be in top physical form, and Glenn liked to jog on Cocoa Beach near the NASA Florida training facility.

NASA

John Glenn was the back-up pilot for the Mercury-Redstone Liberty Bell mission and did pre-flight tests to make sure all systems were functioning properly.

other one off the hill down there. I think we stand on the verge of something as big and as expansive as that was fifty years ago.

"I also agree wholeheartedly with Gus [Grissom] here. I think we are very fortunate that we have, should we say, been blessed with the talents that have been picked for something like this. I think we would be almost remiss in our duty if we didn't make full use of our talents. Every one of us would feel guilty, I think, if we didn't make the fullest use of our talents in volunteering for something that is as important as this is to our country and the world in general right now."

This logic of bigger-than-me purpose, coupled with the fifty-fifty deal he had brokered with fate, would return to Glenn's speeches in the summer of 1998 when the oldest astronaut in the Mercury Program would seek to become the oldest astronaut ever, at the age of seventy-seven.

The Mercury Seven moved to Langley, Virginia, where NASA had its research center. As he had before, Glenn left his family behind so he could concentrate on the task ahead. He stayed in military-style barracks in Langley and commuted on weekends back to Arlington in a teensy two-seat car. Mrs. Glenn had the station wagon.

Each of the astronauts was assigned a different element of the project, to work with engineers on how best to build the hardware needed. Glenn worked on the cockpit layout and instrumentation, Carpenter the communication and navigational equipment, Cooper the Redstone rocket system and emergency procedures, Grissom the hand control and autopilot systems, Schirra the environmental systems and pressure suits, Shepard recovery operations, and Slayton the Atlas rocket and flight procedures. Regularly, the seven men met to brief one another on the progress in their area of specialty. They took trips to a planetarium in North Carolina to learn celestial navigation and to become trained to spot comets, and they went to Arizona for desert survival training. They bounced off Florida's coast in capsule trainers, learning to swim 1,000 feet underwater in their space suits.

Despite the giddy excitement of the astronaut announcement and the flurry of activity that followed, the first two years were not easy for those in Project Mercury. Public thirst for information about the glamorous young astronauts further irritated their brethren in the experimental pilot world, even the pilots flying for NASA.

"Kind of a complicated story, but the fact is they resented all the publicity and fanfare to these people who hadn't done anything yet," Donlan said. "And here [the experimental pilots] are, the elite test pilots of the country, maybe the world out there. Unfortunately most of them were working under secret programs so their heroism was hidden from public view."

But public attention is inseparable from scrutiny. In October 1959, *Newsweek* delivered a broad lashing to the Mercury Program, which was now almost a year old even though the astronauts had only been named seven months earlier, by publishing an article entitled "How to Lose the Space Race." It summarized its view of NASA's situation by answering the challenge raised in the title: "Start late, downgrade Russian feats, fragment authority, pinch pennies, think small, shirk decisions." Then the *New York Times* did its own story months later in which it dismissed the scientific merit of the small satellite launches that Eisenhower had begun to trumpet. "It is not good enough to say that we have counted more free electrons in the ionosphere than the Russians have. . . . We must achieve the obvious and the spectacular, as well as the erudite and the obscure."[6]

All of this heightened the need for a public relations victory that was very long in coming. The development of a reliable launch vehicle,

Each Mercury astronaut was assigned to work with engineers on how best to build the spacecraft's hardware. Glenn worked on cockpit layout and instrumentation.

a rocket, proved difficult. Both of the U.S.'s primary rockets, the small Redstone and the big Atlas, were having too many problems to achieve a "man-rating," or a record of success that would allow consideration of actually strapping a human being to one of those things.[7] Fiery failures seemed to dominate the news out of NASA through 1960 with one spectacular rocket explosion after another. All on film, all covered in the press. The Soviets may have had similar problems, but no one saw them. (In fact, they lost one capsule in orbit because its retrorockets failed, and the thing circled the Earth uselessly for four years.) Mercury seemed to be alone in the struggle to reach space, trailing badly the powerful effort of the Soviets. And 1960 was a presidential election year. Space exploration fit perfectly into the new-frontier campaign themes of the youthful candidate John F. Kennedy, and although he didn't often campaign on space directly, the implications to

NASA's work were clear. It was here that the "missile gap" with the Soviets grew out of proportion with reality, and here that the prospect of achieving human spaceflight behind the Russians was made most painful. Vice President Richard Nixon was saddled with the Eisenhower Administration's reserved response to Sputnik in comparison to Kennedy's "space must be our destiny" rhetoric.

"We are in a strategic space race with the Russians, and we have been losing," Kennedy wrote in answer to a solicitation of views by editors of *Missiles and Rockets* magazine. "If a man orbits Earth this year his name will be Ivan."[8]

Make it Yuri.

And actually, it wasn't in 1960, it was 1961. But the point was the same. On April 12, 1961, cosmonaut Yuri A. Gagarin rode Vostok 1 into a single orbit around the Earth, a feat America wouldn't outdo for nearly a full year.

"We were dead set, we were going to come out ahead in the Space Race, which it was labeled very early on," Glenn remembers. "And so . . . we were mightily disappointed, the whole group of seven of us, when Gagarin went up. Beat us into orbit."

At Langley, what once had been dubbed the "seven-sided coin" that would be flipped to see who got to fly first had landed on Alan Shepard. The bold, brash Navy commander would make a suborbital spaceflight, arcing up and then falling straight back down, passing through the realm of weightless space for only a few minutes. NASA had tried to conceal the identity of its first flying astronaut, and in March had said only that Glenn, Shepard, and Grissom would make the first three flights, not necessarily in that order. Such caution felt more like a tease than anything else, and it precipitated a round of informal bets and rampant guessing over who

would be number one. Glenn took the early lead in newspaper speculation, so much so that it even was suggested that maybe someone had leaked his name specifically to embarrass NASA into choosing someone else.[9] But it was Shepard all along. Glenn, who was chosen to be Shepard's backup, had done a fine job of publicly deflecting the who-goes-first issue by always emphasizing the teamwork necessary for every flight. But privately, his friend Tom Miller says, losing out to Shepard was a blow.

"That upset him more than any time I've ever seen him upset," Miller said.

He wouldn't go water-skiing (a favorite weekend pastime), and he withdrew from those around him, as if he was embarrassed for having not been chosen, Miller said. The two had a confrontation in their adjoined front yards one day during this dark period as Miller tried to get Glenn to realize bigger and better things awaited him. "I think it was probably the harshest words we've ever had," Miller said.

But the echo of Gagarin's eighty-nine minutes in orbit drowned out most everything else. Even when everything was working, Shepard's suborbital trip was only going to cover about 300 miles, yielding maybe fifteen minutes of weightlessness. Gagarin flew 24,800 miles during his hour and a half in orbit, skipping entirely the baby step planned for Shepard. Like Sputnik, Gagarin's flight was an unexpected leap ahead for the Soviet Union.

"All at once the Russians orbited. BANG! It was their first crack out of the box," Glenn said. "And so there was lots of questioning back then, but we just kept plugging along."

And those questions were coming from Kennedy, who had been elected president. Now, it was his Space Race. Within days of the Soviet coup, Kennedy had met with his science and space advisors, including Vice

Glenn (right) served as an observer at the capsule communicator console during Gus Grissom's July 19, 1961, launch. Alan Shepard was the capsule communicator.

President Lyndon Johnson, whom Kennedy had made the White House front man on the space program because of the expertise he had developed as the Texas senator who chaired the Space Committee. The result of the meetings was a memo from Kennedy to Johnson that was designed to allow Johnson a forum for outlining his most ambitious vision for a space program.[10]

In his one-page memo, Kennedy was brief. Just a short list of questions:

• Do we have a chance of beating the Soviets by putting a laboratory in space, or by a trip around the moon, or by a rocket to land on the moon, or by a rocket to go to the moon and back with a man? Is there any other space program which promises dramatic results in which we could win?[11]

- Are we working twenty-four hours a day on existing programs? If not, why not?

- Are we making maximum effort? Are we achieving necessary results?

Johnson responded a week later with a frank memo that answered "no" to all those questions. Johnson then went on for more than five pages to explain how the Soviets' early emphasis on bigger rockets had given them their current success; how the U.S. had an edge

in developing communication, weather, and other science-related satellites that should be better exploited; and how going to the moon was a realistic goal that might even be achievable ahead of the Soviet Union.[12]

"Manned exploration to the moon, for example, is not only an achievement with great propaganda value, but it is essential as an objective whether or not we are the first in its accomplishment — and we may be able to be first," Johnson wrote.

Johnson also sounded off on the inherent need for America to lead in space.

"This country should be realistic and recognize that other nations, regardless of their appreciation of our idealistic values, will tend to align themselves with the country which they believe will be the world leader — the

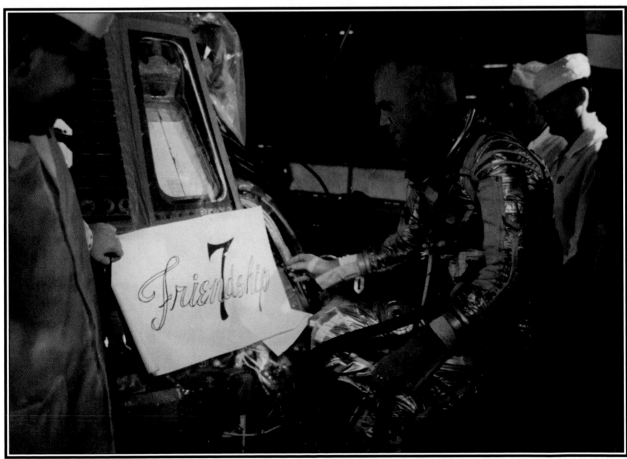

Each astronaut had the opportunity to name his space capsule, and Glenn and his family picked "Friendship 7."

winner in the long run. Dramatic accomplishments in space are being increasingly identified as a major indicator of world leadership," Johnson said.

Johnson's memo, the meetings with space and science advisors, Kennedy's own appetite to generate provocative and positive news after the embarrassment of both Gagarin's launch and the failed Bay of Pigs Invasion, all seem to have conspired to produce the announcement he made on May 25, 1961. In what is sometimes called his Second Inaugural Address, President Kennedy proposed that the nation commit to having an American walk on the moon and come back to Earth before the end of the decade.

Just a few weeks earlier, Mercury had at last put the first American into space. Shepard's May 5 flight into the fringes of weightlessness was declared an "unqualified success" by NASA engineers and scientists, and President Kennedy had watched the thrilling spectacle on television himself. He found it so exciting that he made an unscheduled radio phone call to the Navy recovery vessel to congratulate Shepard who, since he didn't know the president would be calling, had to be summoned to the bridge rather hurriedly.[13] The enthusiasm for Shepard's flight naturally was tempered by the long shadow of Gagarin's recent orbital trip, but the openness with which NASA had conducted the whole affair, live on television, gave the flight a greater boost than the feat itself might otherwise have generated. In fact, watching Shepard from start to finish led some to question whether Gagarin's flight, conducted in secret and shown on a heavily edited film later, had really happened at all.[14]

Heading into July, John Glenn was still a bridesmaid. He'd been Shepard's backup in May — training with him through the centrifuge runs, sitting with him during his launch day breakfast of filet mignon wrapped in bacon, helping with the suiting up. To ease the tension, Glenn the team player had slipped up the gantry and taped a small, amusing picture onto Shepard's control panel for him to see when he got in the capsule (though Glenn steadfastly refuses to reveal what the picture was). Glenn did everything Shepard did except get in the capsule and make the historic ride. For Gus Grissom's launch, which would be identical to Shepard's trip, Glenn was the back-up again. He did all the same things, and he quietly carried the burden of disappointment for having to wait while others went first. He was scheduled to be pilot number three in this phase of the Mercury program, the third and final astronaut to do a suborbital flight using the small Redstone rockets.

But then a couple of things happened. One was that Shepard and Grissom had very successful flights, even if Grissom's escape hatch had opened prematurely while he bobbed in an Atlantic swell and had to scurry out of a sinking capsule. The two flights had done what NASA had hoped — taught them that humans could function in weightlessness, at least for a few minutes, and that astronauts could indeed control a capsule in space, at least for a few minutes. The other thing that happened, naturally, occurred in the Soviet Union. Once again, the Russian space program had upped the ante with a second launch into orbit. While NASA considered whether a third Redstone flight was necessary, Russian jet pilot Gherman Titov spent twenty-five hours in orbit beginning August 6, reporting that he was able to eat and sleep in space. This news proved to be the push NASA needed to skip the last Redstone flight and move on to the second phase of Project Mercury, putting a man in orbit. The bigger, more powerful Atlas rocket would be used for this, and the pilot would be the redheaded guy who was next in line.

Fate was stepping in again. Or dumb luck.

After two successes, NASA decided to fly Glenn on the larger, more powerful Atlas rocket.

NASA

Who would remember the day a Marine lieutenant colonel from a small Ohio town became the third American to pass briefly through space, especially in a summer of pioneering exploration that began with a Soviet circling the globe? Who would notice? Nobody. Glenn would have done his job, that job, and done it well, no doubt, but fate had just intervened to change his assignment. He would not be asked to take a forgettable, incremental step in the long climb to the moon but rather a monumental leap that might herald America's surge past the Soviets toward superiority in space travel.

Glenn's fifty-fifty deal with providence was up again. He had worked himself into close proximity to his country's next great space feat, and the higher power had pushed obstacles aside to deliver him to the greatest one yet. If he did his part, again, he would be the first American to orbit the Earth.

NOTES

1. NASA Project A, Announcement No. 1, December 22, 1958, Invitation to Apply for Position of Research Astronaut-Candidate, NASA History Office, Washington, DC.

2. *This New Ocean*, p. 131.

3. Ibid., p. 161.

4. NASA News Release, May 12, 1959, Sample Questions From Project Mercury Tests, NASA History Office, Washington, DC.

5. NASA Transcript of press conference announcing the Mercury Astronaut Team, April 9, 1959, NASA History Office, Washington, DC.

6. *This New Ocean*, p. 281.

7. Ibid., p. 347.

8. Ibid., p. 284.

9. Ibid., p. 349.

10. Walter McDougall, *The Heavens and the Earth: A Political History of the Space Age* (Johns Hopkins University Press: Baltimore, 1985), p. 319.

11. Memorandum for the Vice President, The White House, April 20, 1961, NASA History Office, Washington, DC.

12. Memorandum for the President, Office of the Vice President, April 28, 1961, NASA History Office, Washington, DC.

13. *This New Ocean*, p. 357.

14. *This New Ocean*, p. 361.

Glenn was involved in pre-flight preparations for his scheduled January 21, 1962, launch — a launch that was postponed when technical problems arose during the pre-launch tests.

CHAPTER 6
THE HERO

February 20, 1962

The clouds over Florida's Cape Canaveral finally parted early in the morning. Finally. Engineers and astronauts, including Lt. Col. John Glenn himself, nearly burst in anticipation of this moment — the liftoff for America's first manned mission into orbit.

The launch should have been in December. And then, when delays consumed December, it was supposed to be January, at the latest. NASA's people knew they could do this, rocket a man into orbit and bring him back to Earth unharmed. They had sent an instrument-filled capsule, and then Enos the chimp, who survived nicely. But the Soviet Union had been into orbit with a man twice now, which caused no shortage of anxiety the world over, and here the Americans sat with the rocket and the capsule and the astronaut, scheduled for launch over and over and unable to get off the ground.

The weather was driving NASA crazy. One day storms would roll in over the launch site, then the next day squalls would kick up a nasty sea in the recovery areas of the Atlantic Ocean where Glenn was scheduled to plop back to Earth for retrieval by a Navy ship. And when it wasn't the weather, it was technical trouble, like the fuel leak discovered in the Atlas booster tank during preflight pro-

cedures that nixed a late January blast-off. Closing in on launch days always made it maddeningly clear that Project Mercury, hoping to be the beating heart of so much spirit and pride and honor, was at its own heart just a million tiny details. Nothing happened when any single one of them went awry.

Throngs of people, news reporters, and patriots flooded the beach south of Cape Canaveral, retreating and returning as launches were postponed and rescheduled. Eventually, the repeated delays wore them down, but instead of giving up on attending the launch, many abandoned the idea of going home. With recreational vehicles and sleeping bags, they covered the beachfront with squatter camps that evolved in some places into tiny municipalities with community rules and hierarchies. One of these towns had its own "mayor" and regularly scheduled post office runs. For everyone everywhere, the stop and go of the John Glenn space shot was wearying.

But Glenn, ever the test pilot, calm in a tough situation, refused to break his deal with fate. He refused to jeopardize the opportunity destiny gave him by getting squirrely now.

"This mission has been in preparation for a long time," he said after the February 15 launch postponement. "I can't get particularly shook up about a couple days' delay. As a

NASA

After he shaved and showered, Glenn ate a pre-launch breakfast of steak and eggs.

signs of blue. The Atlantic north toward Bermuda looked perfect. No mechanical trouble was reported anywhere. The pilot felt good. For the moment, in fact, there were no problems at all. But only for the moment. The truth is a million tiny details can't possibly be tamed on any one day, so the first American in orbit would get there on a winter morning composed not of perfection but rather of a dozen deliberate compromises, a fistful of calculated concessions. And only a few of them would seem tiny.

Glenn rose around 2:30 a.m. A shower and shave braced him before a hefty breakfast of steak and scrambled eggs with toast, orange juice, and coffee. He endured yet another physical exam before arriving at a hangar where he suited up. Minutes after 6:00, Glenn strapped into the Mercury capsule he and his family had named — in the midst of the U.S.-Soviet space rivalry — Friendship 7. As Glenn shimmied into position, he noticed that the respiration sensor on his helmet's microphone had been moved out of position. Fixing it would require burrowing into Glenn's pressure suit, so doctors agreed to live without the monitor on Glenn's breath. Here was the first lost detail. Then an hour later, while the seventy hatch bolts were being wrung home, technicians discovered that one of them was broken. The hatch had to be removed and reseated, and the morning's countdown toward launch froze for forty minutes. Detail number two, blown. As the last bolts were tightened, Glenn, still wary of a betrayal by nature, looked through the capsule's periscope

matter of fact, I'm so happy to have been chosen to be the pilot for this mission that I'm not about to get panicky over these delays. I learned very early in the flight test business that you have to control your emotions — you don't let these kinds of things throw you or affect your ability to perform the mission."

In the on-base newsletter at NASA's Langley facility, Glenn sounded firmly optimistic through the parade of launch holds. "I don't think we should fly until all elements of the mission are ready. When we have completed all of our tests satisfactorily, then we'll go," he said.

And here it was. At long last, the gray cloud ceiling over Cape Canaveral showed to the world outside. "Looks like the weather is breaking up," he said, pleased.[1]

Glenn was snug in the couch, as they called it, though it was like no couch you've ever seen. It was a tiny thing, scarcely a chair, built with room for John Glenn's hips in a space suit, and nothing more. He faced skyward, peering out the window at a bluing sky, then into a mirror that bent his view to the coast below. NASA patched a telephone call through to Annie back home. The conversation ended in the small, comforting ritual John and Annie had worked up through years of harrowing combat missions.

He said, "Well, I'm going down to the corner store and buy some chewing gum."

Then she said, "Well, don't take too long."

Annie didn't cry until she was off the phone.[2]

Liquid oxygen, pure flammability, streamed into the massive tanks beneath Glenn. The capsule began to vibrate and shudder as the thin metal of the tanks stretched like a balloon to take the liquid. Some at NASA didn't want to use the Atlas at all, for this very reason. Too flimsy. Too dangerous. They had nicknamed it the "steel balloon" and pressed to keep the Redstone rocket for all manned flights. They were smaller but more heartily constructed, and had put Alan Shepard and Gus Grissom up in their flawless suborbital flights. The rocket debate didn't end until an engineer took a sledgehammer to a pressurized Atlas fuel tank and proved, beyond much doubt, it

NASA

Glenn awoke at 2:30 a.m. on February 20, 1962, to get ready for his launch.

was strong enough to hold together.[3]

Besides, getting into orbit called for the big guns.

The metal tower holding the Atlas rocket in place at last pulled away. A real launch. By now Glenn had been lying still in his seat for nearly four hours. He wiggled in place and felt the ninety-five-foot missile rock softly on the concrete pad.[4]

At 9:47 a.m., the big gun fired. "Godspeed, John Glenn," said Scott Carpenter, Glenn's back-up for the flight.

NASA

Glenn left Hanger S, strode to the launchpad, and was strapped into his seat shortly after 6 a.m. — and waited nearly four hours before liftoff.

Ten.

Nine.

Eight.

Seven.

Six.

Five.

Four.

Three.

Two.

One.

A throaty rumble starting at the ground climbed forcefully up the rocket, enveloping Glenn and his couch in a blurring shake.

"Roger," Glenn responded, his voice tight, staccato. "The clock is operating, we're underway."

Up.

"Reading you loud and clear."

"Roger," Glenn said. "OK, a little bumpy long about here."

"Roger."

Traffic in New York City clogged to a halt. Commuters in Grand Central Station, thousands of bodies with thousands of destinations, solidified into a single audience rapt below the giant television, a twelve-foot by sixteen-foot wonder of the day tuned live to the launch.[5] In the city's subway, tin-sounding loudspeakers blurted the news: "Col. John H. Glenn Jr. has just taken off in his rocket for orbit. Please say a prayer for him."[6]

It was the same all over. One hundred thir-

NASA

At 9:47 a.m. on February 20, 1962, the Atlas rockets fired, the capsule began to vibrate, and Scott Carpenter said: "Godspeed, John Glenn."

ty-five million Americans at TV sets, not to mention radio, the rest of the world, and the three TVs in the Glenn home in Arlington. Each Glenn TV was tuned to a different network broadcast, the volume cranked on one. A crowd had assembled on the Glenn's lawn, spilling over, of course, onto his best friend Miller's lawn. It was cold, and the Millers, between tending to Annie and the kids during the nerve-wracking day, turned their house over to the press corps, brewing up coffee and giving them shelter.

At the White House, President Kennedy caught every minute of the coverage from an array of televisions arranged throughout the residence and executive wing. In places that didn't have a TV — the Oval Office, the dining room where Kennedy had a breakfast meeting scheduled with legislative leaders — sets were brought in and hooked up temporarily. At liftoff, the president was too tense to sit down. He stayed on a direct phone line to Cape Canaveral through the first several minutes of the launch, listening for himself as mission control reported everything was going fine.[7]

And Glenn roared on. The primary Atlas booster fed the last of its fierce-burning oxygen to the engine, and as the pressure eased in the fuel tank, the "balloon" began to flex. Hurtling spaceward, Glenn felt like a man out

NASA

Glenn fit snugly inside the tiny capsule, sitting in a chair that had barely enough room for him. But from a window he could see the beauty of space.

on the end of a springboard.

Three small jets kicked to life, thrusting the empty rocket clear of the capsule. Glenn didn't expect to feel the push, because those jets are so small, but he did. The automatic capsule control system, a series of tiny jets all over the capsule designed to generate thrust and turn the cone-shaped module when necessary, whirred. The capsule spun around, and Glenn could see the Atlantic Ocean and the spent Atlas falling away.

"Zero G and I feel fine!" Glenn radioed, just five minutes after lifting off. "The capsule is turning around. Oh, that view is TREMENDOUS."

"Roger 7," the cape barked back. "You have a go, at least seven orbits."

"Roger, understand go for at least seven orbits."

They weren't going for seven orbits, but Glenn was thrilled to hear they had the speed and trajectory to complete their three circles with room to spare. He loosened his chest strap. Weightlessness felt good, easing his body free of the cramped couch, erasing the pressure. "This is very comfortable at zero G. I have, uh, nothing but very fine feeling. It just feels, uh, very normal and very good," he said.

The purpose of this flight was so very simple, in contrast to the dense complexity of the effort to carry it off. They just wanted to know for sure that it could be done. Scientists mostly wanted the mystery solved, the basic questions answered about how it is to be in orbit. In space. Glenn's list of tasks was fundamental. A bungee cord to tug and stimulate his heart rate. An eye chart taped to the control panel not two feet from his face. A standing request for blood pressure readings. He shook his head violently, then stopped. Nausea? No, fine. He held a gloved hand out before his face and waved a finger back and forth quickly; his eyes followed the finger tip, which was lighted by a tiny, built-in bulb.

"I have had no ill effects at all from zero G," Glenn radioed. "It's very pleasant, as a matter of fact. Visual acuity is still excellent. No astigmatic effects. Head movements cause no nausea or discomfort, whatsoever."

On a highway, you get a sense of speed. Sixty miles an hour, and you know it. In space, the sensation of speed is suspended. Glenn couldn't tell he was moving at 17,000 m.p.h. Just forty-three minutes into his flight, about 10:30 in the Florida morning he'd just left, Glenn's capsule circled around to night. He was awed by the brilliance of colors, the beauty of a fast-closing sunset 100 miles above the Earth. Here, the horizon is not where the Earth meets the blue sky, but where the Earth meets limitless black space.

"That was about the shortest day I've ever run into," Glenn said.

It was nighttime in Australia as Glenn went by, and he saw the city of Perth burn brightly below him. Street lights glowed and people in the coastal city left porch lights going all night as a welcome for Glenn's pass. Some staked white sheets across their lawns to reflect the light. Taxi drivers pulling graveyard shifts flickered their headlights when radio news told them Glenn would be overhead.

"The lights show up very well," Glenn radioed. "Thank everybody for turning them on, will you?"

Glenn flipped the clear plastic face shield back and squirted apple sauce from a tube into his mouth. He wasn't hungry, just another test. "Have no trouble at all eating, very good," he told mission control.

Crossing back into day, completing the first of three ninety-minute Earth orbits, Glenn's capsule suddenly was surrounded by a swarm of lighted flecks. Little stars, he told the cape. It was like a shower of tiny stars. NASA didn't have a clue what they could be. Neither did Glenn. They disappeared with the sun. But Glenn couldn't ponder them long.

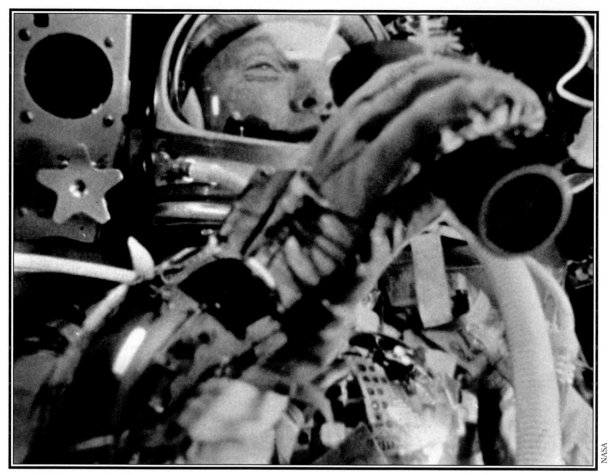

NASA

Glenn flipped the clear plastic face shield back and squirted apple sauce from a tube into his mouth. He wasn't hungry, just another test.

Soon after his second "day" began, another detail, this one crucial, spun out of control. The small jets on one side of the capsule's automatic control system stopped working. They started again, but then the other side went on the blink. Without them, Friendship 7 began drifting to one side, like a car with bad alignment.

Glenn switched the control system to manual, and the capsule responded perfectly to its pilot. This was no longer a crisis, and Glenn felt good at the helm. It was satisfying to be needed, to be more than "spam in a can." But driving by hand used more fuel than the automated system, and it would be deadly to run out of fuel. Glenn would have to be very careful. The capsule could return to Earth only by reentering the atmosphere facing a particular way, heat-shield down, and

you can't get the capsule into position without those jets. And this was not the kind of trip where NASA could afford to stock the capsule with a lot of surplus anything.

On the ground, Cape Canaveral fretted over the control problem. But soon, something more serious demanded their attention. This was no pesky detail, either. One of the engineers got an ominous reading from the telemetry control console. According to the console, Glenn's heat shield — the special fiberglass and resin belly of the capsule that keeps the space module and its astronaut from burning to dust on reentry — had mysteriously popped loose.

The shield was designed to detach. Packed beneath it was a heavy rubber landing bag that looks like a giant accordion. The idea is that once the capsule is back in Earth atmosphere and floating by parachute to an ocean splash landing, the heat shield pops free, stretching the landing bag down into a shock absorber. But this? Coming unbuckled in space? No. This was fatal. The landing bag would melt when the capsule reentered Earth's atmosphere, the heat shield would slip out of position, and Glenn and his couch and everything else in Friendship 7 would incinerate into infinite radio static.

On the ground, NASA moved quickly. Almost halfway through the flight, they called the eighteen monitoring stations around the globe and asked each of them to delicately ask Glenn to double-check his own control panel and make sure the landing bag deploy switch was OFF. Don't tell him why we're asking, NASA said.

Now, Glenn is a smart guy. All those weeks of training on the procedures, all those hours of repetitive simulations, he knew this was out of the ordinary. Something was wrong, but no one would tell him exactly what. "Did someone report that the landing bag could be down?" Glenn asked, after another request to check it, three hours into his flight.

"Negative," came the reply. "We had a request to monitor this and to ask if you heard any flapping."

So he kept working. With the loss of automated controls, Glenn had to scrap a host of experiments, pass on another squirt of food because he was too busy steering the capsule. He kept checking the autopilot to see if it had returned to normal. On his second sunset, Glenn turned the capsule to see if he could get a better view of the luminous swarming stars. He blipped the thrust on one of the control jets to see if condensation from them was causing the tiny star swarm, but that did nothing.

At the cape, NASA engineers fought. Some wanted to ignore the indicator light that showed the heat shield problem, since no other back-up indicators gave the same read and the capsule's own system showed everything was fine. But what if? The flight planners called their top capsule design man in Houston and asked his opinion on a major change in the reentry plan.

Coming back out of orbit is no easy thing. The only way out is to slow down, to break the perfect balance between centrifugal force and gravitational force that is the essence of being in orbit. Doing this at 17,000 m.p.h. requires more rockets. The Friendship 7 had a bundle of three rockets slung under its belly, across the heat shield, and held in place with three metal straps. To begin the return to Earth, these retro rockets fire, throwing a burst of power in the direction opposite the orbital path. When the capsule slows down, gravity overpowers the centrifugal force and begins pulling the capsule back to Earth. Ideally, the retro pack, as it's called, is released after stopping the capsule and drifts off into space, so the heat shield is smooth and clear to deflect the phenomenal heat generated by leaving the vacuum of space and slamming into millions of densely packed air molecules in the Earth's atmosphere.

But NASA had another idea today. They asked the designer in Houston if the capsule could survive reentry with the retropack in place. Would it disrupt the aerodynamics of the capsule? If that happened, and the capsule tipped over during reentry, Glenn would die. NASA hoped the metal straps of the retro pack would hold the heat shield in place long enough for sheer wind force to take over. The Houston designer said he'd tested that very scenario in a wind tunnel the year before. The capsule will handle it fine, he said.[8]

Some at NASA thought this was a lousy idea, including flight director Christopher Kraft. He thought it was more risky than the chances that a lone indicator light had detected a problem not picked up by any other monitors. The debate went on, but it didn't last long. It couldn't.

With about five minutes left before the scheduled reentry, NASA radioed Glenn.

"This is Cape Com, Friendship 7. We are recommending that you leave the retro package on through the entire reentry."

"Uh, this is Friendship 7," Glenn responded. "What is the reason for this? Do you have any reason?"

"Not at this time," the cape said. "This is the judgment of Cape Flight."

A public address system at Cape Canaveral gave minute-by-minute updates to reporters and broadcast crews throughout the flight, sometimes even patching through portions of Glenn's gravelly radio conversations with mission control.[9] But when engineers found the heat shield problem, the updates only made veiled reference to the issue so no one knew how chilling the situation really was.

Glenn was always big on keeping the family involved in his work, explaining all the technical details to put everyone at ease. He had brought home a model of the Friendship 7 for the family to see, since NASA wasn't letting family members on the cape. And he outlined details of the flight so thoroughly that during the flight, Annie and their teenage son, David, used stopwatches to time his orbits so they could track where he was throughout the day.[10]

"He would sometimes spend an hour or so in the evening," remembers Tom Miller, "just sitting around in their living room explaining how things worked and so forth."

So on flight day, even a veiled reference to the heat shield would have come through loud and clear to Annie. But she didn't have to wait for television to let her know, because NASA called. Suddenly, all that information Glenn had given, all those details that had seemed reassuring when she first heard them,

became way too much information. She would not misunderstand it for what it meant for the capsule to lose its heat shield. "If you were looking at [Annie] you couldn't tell it registered," said Miller, who was in the house and looking at Annie. "But she knew."

Glenn had figured it out, too.

Ultimately, NASA told Glenn. Just in time for him to cinch back his chest strap and wait to see if the explosive heat of reentering the Earth's atmosphere would come through the capsule and eat him alive. Moments after the first heat pulse began glowing outside, Glenn saw a strap from the retro pack slap against the window and flame off into space.

"Uh, this is Friendship 7, I think the, uh, pack just let go," Glenn reported.

Static.

Chunks of flaming debris streaked by his window, and Glenn, convinced the retro pack already was gone, thought his heat shield was burning away. It was supposed to flake off with the heat, but not like this. Not whole chunks. He waited, thinking he may soon start feeling the fire against his back.

At the cape, to the world, in the house in Arlington, Glenn was lost to all communication. The heat surrounding his capsule blocked out radio contact. Monitors stopped registering. This was expected, and it wasn't long, but that didn't make it easy.

Static.

As difficult as the pressure of launch day can be, this, this was that multiplied by ten. Maybe 100. A single instant, Glenn either coming back on the radio now or not, would decide so much. Sure, the space program might survive an astronaut's death, but how? After what the Soviets had done? We save the chimp but kill a man? Like it or not, this single instant in which John Glenn either was there or he wasn't had to mean everything. Either way. The weight of Project Mercury, and the fledgling agency called NASA, was on

NASA

While the Soviets preferred reentry of their spacecraft on land, the United States preferred landings by sea, where astronauts were picked up and taken to a nearby ship. Glenn was picked up by the destroyer USS Noa *in the Pacific Ocean.*

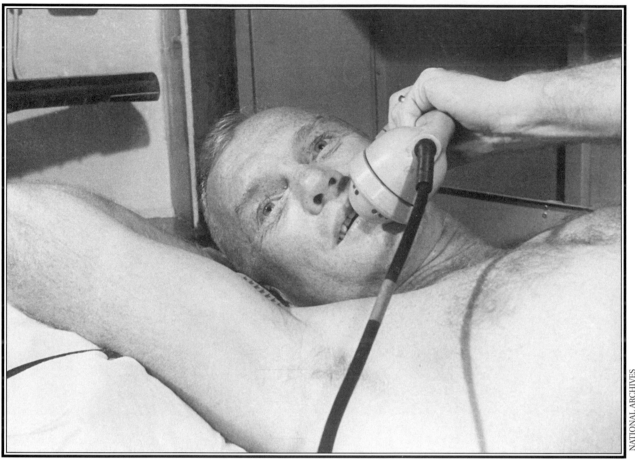

NATIONAL ARCHIVES

After four hours and fifty-five minutes in orbit, astronaut John H. Glenn Jr. splashed down and was brought aboard the destroyer USS Noa, *where he spoke with President John F. Kennedy.*

the head of a pin.

And then, through the atmosphere, the static, the flames: "That was a real fireball, boy!" Glenn's voice at last burst through.

But the elation of Glenn's survival through reentry, like the faulty autopilot trouble that arose after the first orbit, was soon overshadowed by a bigger problem. Another rogue detail lost to chaos. Coming out of the heavy G forces of reentry had set the capsule swinging violently. Too violently, in fact, for Glenn to control even with the manual pilot system. He thought the Friendship 7 might swing all the way over and fall to Earth upside down, with the capsule's bottle-top end striking the ocean first. No shock absorber was built into that end, and from 80,000 feet, this would be as deadly as any rocket fuel explosion or heat shield slip.

"Rocking quite a bit," Glenn radioed. "I can't damp it either."

He was pumping the tiny jets of the control system, still hoping the thrust might deaden his swinging, when the last dangerous detail slipped away. The fuel. The control system jets went dry. All that manual flying caught up with him, and what little power Glenn had to slow the frightening swinging was gone. He told the cape he wanted to deploy the preliminary parachute immediately, maybe settling things down. He was still a bit higher than he should be to pop that chute, but if the capsule flipped, then opening the stabilizing chute might not be enough to right it.

But just as Glenn reached for the initial parachute switch, it popped automatically. The capsule slowed its violent swing, and one minute later the main parachute blos-

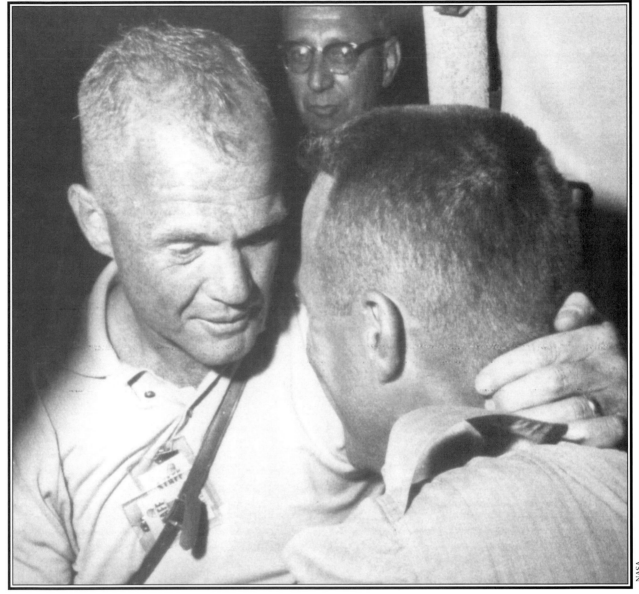

NASA

Glenn always talked about the importance of the Mercury team, one of whom was Scott Carpenter, the back-up for the Atlas flight, who greeted Glenn a day after his return.

somed overhead.

"Rate of descent has gone to about forty-two feet per second," Glenn said. "Chute looks very good."

The searing reentry heat warmed the air inside the capsule. As he drifted to a spot within a mile of a Navy destroyer, Glenn began to sweat. Heavily. His body temperature rose a full degree, and five pounds of water weight dissipated from his lean frame. The capsule temperature was over 100 degrees inside as Glenn bobbed in the Atlantic Ocean about

800 miles from Bermuda. Maybe it was the dehydration, or perhaps the heat, or quite possibly just Glenn's own vulnerability to seasickness — anyone who saw him on a Navy ship knew about that — but for the first time during this brave day, Glenn didn't feel so good. His stomach was queasy.[11]

Through the radio, Glenn and the recovery team blurted back and forth. A crewman told Glenn he'd be aboard ship in two minutes, and Glenn, suffering from the heat, blipped out a calm, "Roger." But that heat!

Glenn was greeted by his wife, Annie, and Vice President Lyndon B. Johnson upon his return to Patrick Air Force Base from his Mercury flight.

Sweat bathed him. He went on, understated, intense: "This is Friendship. I'm very warm. I'm just remaining motionless here trying to keep as cool as possible. I'm extremely warm at the moment."

Making matters only slightly worse, the Navy's winch crew banged the capsule against the ship's steel hull as they hauled Glenn aboard. Just over five hours after launching, the Friendship 7 capsule at last came to rest on the deck of the destroyer *Noa*. Inside, Glenn boiled. He was supposed to climb up through the hatch in the top, but between the heat and his stomach,

Glenn just wanted out. He radioed the ship's bridge and asked that everyone be cleared from the deck, away from the capsule. With a slap at an emergency lever, a set of small explosives blew the side hatch off and onto the deck. Then the last lost detail: The emergency lever recoiled and put a scratch through Glenn's glove, biting the skin of his knuckles.

"It was hot in there!" he said, out at last.

A sailor handed him a glass of iced tea.

NATIONAL ARCHIVES

John Glenn (with his wife and Vice President Lyndon B. Johnson) had insisted that all seven Mercury astronauts be included in the New York City parade that marked his return because he wanted his colleagues to share in the celebration.

None of the previous events in Glenn's life, or in Project Mercury's either, for that matter, even hinted at the adulation to come. Turns out the heart of American spirit, pride, and honor that Project Mercury hoped to capture with a space triumph was beating the whole time in the chest of the redhead from Ohio. His handsome features, his family, his persistent humility in talking about the "team" that had made his orbital flight succeed, created an instant celebrity where once had been only a passion for duty. He called himself a "figurehead" in the vast effort behind his flight, and during the first press conference, after wading through 100,000 people swarming his parade route for autographs, Glenn stopped to thank his backup in the mission, M. Scott Carpenter, by telling how Carpenter recently saved a man while scuba diving.

"That was very impressive," Glenn was saying. "When you have an aqualung and you're down at about eighty-foot depth and you're trying to give your own air to somebody else to help them out, that is almost heroic in my book."[12]

But the hero everyone wanted was Glenn himself. And not just in America, but all over the world. Although Soviet men still had done more in space than America, flown longer in orbit, Glenn's well-executed mission on live television seemed almost to vanquish the Soviet space gap in one fell swoop. Around the world, news reports hailed Glenn's mission as a triumph for freedom itself.

"Glenn's flight carries much political weight," a newspaper in Vienna wrote. "His

John Glenn returned to a hero's welcome, and parades were part of the public celebration. The special moments in his life — like the parade in Washington, D.C., on February 6, 1962 — were shared with his wife, Annie.

victory in space is a victory in the Cold War. . . . Actually, he did not fly merely for Kennedy or the State Department — he flew for mankind." [13]

Soon, the redhead would address a joint session of Congress, submit to a ticker tape parade as only New York City can throw one, and sign the National Geographic Society's Fliers' and Explorers' Globe that previously was autographed by a small fraternity of men including Charles A. Lindbergh, Richard E. Byrd, and Sir Edmund Hillary. At a New York gallery, less than two months after his flight, a few private letters written by Glenn were auctioned off along with the correspondence of Lindbergh and Albert Einstein.

President John F. Kennedy took a genuine liking to America's new hero.

NASA

Glenn's achievement was more than a victory for the United States' space program — some thought it was a victory in the Cold War for President Kennedy.

Newspapers reported that a rare books dealer bought the Glenn letters, in which he dickered with a car dealer over trouble he had with his little import, for $425. It felt then as if a legend might be taking root.

When NASA contemplated a twenty-four-hour orbital mission, which ultimately was scrapped, they dropped the coy game about naming pilots and openly used Glenn's name, according to news reports. But at the same time, something else was happening that was both valuable to Glenn's future and, apparently, deadly to his space career. President Kennedy had taken a real shine to Glenn, inviting him to White House events and out

to the family compound in Palm Beach. Glenn was seen water-skiing with the First Lady (which earned him a telegram from a marine safety expert who scolded him for not wearing a life vest: "Your actions form patterns for many, yet you may have unknowingly risked the lives of many."). But Glenn would not fly in space again because, as he later learned, Kennedy wouldn't let NASA risk the life of the popular American hero with another spaceflight.

The Mercury Program flew three other orbital missions over the next year, concluding with twenty-two laps around the Earth by Gordon Cooper, but none of those flights

The first American to orbit the Earth was in great demand, and John Glenn addressed both houses of Congress after his return from space. With Glenn on the steps of the Capitol — a place he would soon find routine — are Annie and their children, David, 16, and Lyn, 14. Glenn's parents are behind his wife and daughter.

captured America the way Glenn's did. Months after Mercury had been closed down, having been replaced by the Gemini program, NASA still was trying to get a handle on requests for astronauts to make personal appearances. Requests for Glenn ran at twice the rate of the other six combined.[14]

Fate was doing it again. Raising the ante, giving the fifty percent Glenn would have to match, even before he knew how. Destiny was building in him a new set of talents, for public speaking and popular leadership, while blessing him with incomparable name recognition. And as he had said before about pilot skills and physical condition, Glenn soon would feel an obligation to do with these gifts as much as he could.

Public office would call.

NOTES

1. *This New Ocean*, p. 423.

2. *Life* magazine, Loudon Wainwright, March 2, 1962, p. 31.

3. *This New Ocean*, pp. XX.

4. NASA Report: Results of the First United States Orbital Space Flight, February 20, 1962, Mercury files, National Archives and Records Administration, Southwest Branch, Fort Worth, Texas.

5. *Life* magazine, March 1962.

6. Unidentified newspaper article, February 23, 1962, NASA History Office.

7. Account of White House.

8. Telephone interview with retired NASA engineer, Maxime A. Faget, at his home in Houston, Texas, February 1998.

9. Transcript of Public Address announcement of MA-6 launch, Col. John Powers, NASA History Office.

10. *Life* magazine, Wainwright, p. 31.

11. NASA Postlaunch Memorandum Report for Mercury-Atlas No. 6, March 5, 1962, National Archives and Records Administration, Fort Worth, pps. 7-15.

12. *The New York Times*, February 24, 1962.

13. NASA Free World Media Treatment of First Orbital Flight, March 5, 1962, p. 4.

14. Correspondence. Robert Gilruth, director of Project Mercury, to Dr. George E. Mueller in NASA, November 18, 1963, NASA History Office.

The third time was the charm for Glenn, who finally won the U.S. Senate seat in 1974.

CHAPTER 7
THE POLITICIAN

Unlike the other phases in John Glenn's life, his move into politics was not a seamless transition for which he was ideally prepared. Glittering name recognition and a fascinating life story of ambition, achievement, and virtue is — in the world of electoral politics — nothing but a good start. Winning elections also requires campaigning — kissing babies, shaking hands, rousing speeches. And raising money.

These things were not Glenn's strengths. In the Marine Corps of Glenn's day, working the angles and selling yourself for plum assignments — which was called "sniveling'"and which Glenn did better than most — may have required a gift for self-promotion similar to that of a politician. But selling oneself in the military was a more self-determined enterprise than running for office. Ultimately, the successful sniveler relied primarily on his own drive to achieve, to conquer the assignment he worked feverishly to get. Obsessive attention to detail defeated big-picture concerns, which were uniform and understood and clearly the responsibility of higher pay grades anyway. Politics, on the other hand, is far more open-ended, the field of play far more dynamic. Glenn might not have guessed when he announced his candidacy for the U.S. Senate in 1964 that he had so many lessons still to learn about getting what he

wanted. Or, more to the point, that it would take him a full decade from that moment to finally realize his ambition to represent Ohio in Washington.

In the first year after his orbital fight, Glenn just wanted to go back into space. He hounded the higher-ups at NASA to put him back into the flight rotation, but every time he brought it up they delayed him. Not yet, they said. Let's wait. As time wore on, the denials became more blunt until Glenn finally was told, "Headquarters doesn't want you to do it." They didn't tell him what Glenn learned years later about JFK's intervention.

"To stick around there being the world's oldest . . . used astronaut wasn't my idea of how I wanted to spend the rest of my life," said Glenn, who was forty for his first flight, the oldest of the seven astronauts. "And they were talking to me about taking over some of the management of training and things like that, and I could have done that, but I just, I wanted to do other things. I didn't want to stick around always being a trainer."

At the same time, his relationship with the Kennedy family had bloomed nicely, introducing a fairly obvious route into political service for the enormously popular astronaut. JFK was hungry for Glenn to join the Democratic Party and announce his candidacy for federal office in the fall of 1963 for sev-

The thrill of Mercury quickly turned to frustration. John Glenn wanted to ride in space again, but NASA wouldn't let him.

eral reasons. Kennedy had lost Ohio to Richard Nixon in the 1960 presidential race, and with Kennedy's Civil Rights position likely to cost him dearly in the South he could not afford to lose Ohio again in 1964. Running side-by-side with Glenn would have been ideal. Also, the sitting senator from Ohio, Democrat Stephen M. Young, looked very vulnerable to the expected Republican challenger, U.S. Rep. Robert A. Taft Jr. a member of the famed political family. Kennedy's death in November shocked the American political system, but by early 1964 Glenn emerged officially as a candidate to challenge Sen. Young in the Democratic primary. If Glenn could win it and face Taft in the general elec-

tion, what an exciting race that would be! The Democrats' best new hope against the heir to the Republican family legacy that culminated in William Howard Taft, America's twenty-seventh president!

Then the oddest thing happened. Within days of the second anniversary of his orbital flight, Glenn did a life-altering thing: He took a shower.

Standing in the tub of his Columbus apartment, dripping wet, he noticed that the sliding mirrored door to the medicine cabinet was not on its track properly. He stepped out of the tub and onto a throw rug to lift the heavy door back into place. But the rug slipped, his feet flew out from under him, and

Glenn crashed to the floor, striking his head on the metal track of the tub's own sliding doors. The medicine door came down on top of him, breaking on his head. Glenn's vestibular system, which is located in the inner ear and governs human balance, was badly damaged. Nothing permanent, but the auspicious political career of Col. John Glenn, barely a month old, was halted amid a lengthy hospital stay and persistent, dangerous nausea and vertigo. On March 30, 1964, Glenn acknowledged that he was physically unable to campaign and followed his doctor's orders by withdrawing from the Senate race. "In

ASSOCIATED PRESS

By 1964, John Glenn had decided to try politics and challenged Ohio's incumbent U.S. Sen. Stephen Young, a Democrat. The race looked promising to the astronaut, pictured here with his wife, Annie.

The promise turned to disappointment. Glenn — with his wife nearby — pulled out of the 1964 U.S. Senate primary race because of an injury that disrupted his balance.

announcing my candidacy for the Senate, I wanted to present my views as a candidate," Glenn said in a statement issued from his hospital bed. "I did not want to run just as a well-known name."[1]

What followed for Glenn was a period of extraordinary difficulty. Through the summer of 1964, Glenn recovered slowly from his vicious fall. He had left NASA and retired from the Marine Corps, and now that his run at politics had gone badly he didn't have a job. A friend who telephoned Glenn that summer said later that he "seemed like a guy who thinks his life has come to a dead end."[2] What's more, Glenn's brief Senate campaign had racked up thousands of dollars in expenses that now had to be paid off. Glenn refused to hold postmortem fundraisers to get supporters to help pay those campaign bills

because he considered himself solely responsible for his need to leave the race. Instead, he and Annie sat down and wrote checks from their own accounts until all the bills were paid, and all their reserves were gone. "If there's a war on poverty, where do I go to surrender?" Glenn joked at the time.[3] When he was with NASA, Glenn and the other astronauts had signed an exclusive deal to tell their stories to *Life* magazine — an agreement brokered by NASA itself — and while that $70,000-plus payoff was good, the money was gone. Glenn had been offered plenty of opportunities to cash in on his celebrity, from endorsing cars to producing movies, but he declined them all. Instead, when he was at last well enough to work, he accepted an executive management job with the Royal Crown Cola Company. It paid well and offered lucrative stock options, but what appealed to Glenn was that the job was not in advertising or promotions. RC didn't want to use his face to sell soda pop.[4]

With his life back in motion, Glenn worked diligently in business and in politics. He devoted time to the Democratic Party, developing allies and positioning himself for another run at the Senate. His business demanded extensive travel, and over the next few years he visited dozens of countries, sometimes speaking for NASA but mostly working on behalf of RC to scope out market opportunities abroad. But even as the years piled on between Glenn and his historic space ride, he remained a national figure. In December 1967 he appeared on "The Tonight Show," bantering with host Johnny Carson as easily as any Hollywood leading man.

"I remember almost crying," Carson said, trying to set a mood for the interview, "when you realized that somebody was sitting on top of that rocket in that capsule. You, of course, were probably too busy and having gone through it many times. . . ."

"Are you sure I wasn't crying a little there, too?" Glenn said.

"Will you stop it!" said Carson, laughing. Then, mimicking a parent soothing a baby, asked, "Did you want your bankee? Did you?"

The audience ate it up.

So by 1970, with Sen. Young retiring, Glenn seemed to be a shoo-in to defeat Cleveland millionaire Howard Metzenbaum in the Democratic primary. However, his opponent was a seasoned political operative with strong ties to the party and to organized labor, two essential ingredients for success in Ohio. And just as importantly, Glenn still had much to learn about presenting himself in politics. When the two squared off in Dayton two months before the May primary, Metzenbaum suggested Glenn was inconsis-

tent in his stand on such things as gun control and government subsidies, and the Marine colonel absolutely snapped.

"When you say I go around the state saying one thing in one place and another thing in another, you lie, Howard," Glenn spat. "You're a liar."

The audience applauded but Metzenbaum didn't retreat, and he didn't rattle. "When you said I lie, sir, you lie about that, sir," he responded.

With that, Glenn jumped to his feet, perhaps headed across the stage to settle the issue decisively — audience members feared a fist fight — before the state party chairman grabbed Glenn's sleeve and tugged him back into his chair.[5]

Metzenbaum spent heavily on a television and radio campaign, and Glenn's organization never quite responded. Communication with local newspaper and television operations was poor, leaving many of his events out of the news entirely. His tremendous name recognition at times worked against him, too, making it tough to convince donors that he needed money to get his message across. And finally, news events and party demographics conspired in the days before the election to defeat Glenn again. The U.S. invaded Cambodia the week before the primary vote, and one day before election day, Ohio National Guardsman shot four students to death during demonstrations at Kent State University. Although Glenn opposed the Vietnam War and was eloquent in reaching out to distrustful young Americans, the liberal element of the Democratic Party that controlled the primaries simply didn't warm to the retired Marine colonel, choosing Metzenbaum by just over 13,000 votes. Metzenbaum lost the general election to Robert A. Taft Jr.

Perhaps a bit embarrassed by the defeat, Glenn retreated into business where at last his

Glenn and his wife hit the U.S. Senate campaign trail again in 1970, but this time he was narrowly defeated in the Democratic primary by his rival, Howard Metzenbaum.

financial picture brightened. This time, for instance, he let donations pay the campaign debt, and he began investing his RC salary in hotels. He had one more short brush with poverty in 1973 when his investment in a Holiday Inn in Columbus hit hard times. But with the exception of that one spot, Glenn had put lots of money into hotels — mostly in Florida near the exciting new theme park called Disney World — and was on his way to becoming a wealthy man. Still, though, Glenn wanted to be a senator. He wasn't going to abandon that goal, and he wasn't going to settle for a lesser one. Attempts by Gov. John

Gilligan to recruit Glenn to be his lieutenant governor for the 1974 race were rebuffed rather abruptly by Glenn, who disliked that kind of deal brokering over elected positions. Glenn wanted to be senator, period. At a high level party meeting designed to draw Glenn into reconsidering, he blasted the governor and everyone within earshot for trying to perpetrate a "fraud on the voters" by urging him to run for an office he didn't really want.[6] Glenn went after Metzenbaum again in the party primary for Senate, only this time Metzenbaum's advantages included incumbency. The senator whose seat Glenn wanted,

Republican William Saxbe, had been appointed U.S. Attorney General by President Nixon, which forced Ohio's governor, the man Glenn had just embarrassed, to appoint a replacement to fill the final months of Saxbe's term. He chose Metzenbaum, saying he had based his decision on Metzenbaum's primary victory over Glenn in 1970.

But unlike the previous races, Glenn was prepared for politics this time.

"In 1970, we got a late start, we didn't get the right people, and we had no money," Glenn said at the start of the '74 campaign. "It wasn't that we were so naive that we didn't recognize things were going wrong. It's just that we couldn't do very much about it. This time, we hope to convince people that we are worthy of their contributions."[7]

And this time, Glenn got some breaks. During the course of the campaign, Metzenbaum released financial disclosure forms showing he had paid no income tax in 1969 despite his considerable wealth, and he acknowledged that he was fighting with the IRS over the issue.[8] Glenn released his own records, which showed he had paid more in taxes than his opponent, and at last the personal attributes of John Glenn — his Boy Scout goodness — could be easily translated to the political arena. He didn't have to learn anything new to be the candidate of integrity, and he hammered at Metzenbaum on the issue with military relentlessness. As before, the race was nasty. And ridiculous. Both candidates fended off waves of accusations up to and including the allegation that they lived in houses built with nonunion labor.[9]

Metzenbaum tried fighting back with the theme that Glenn was strictly a government product, hoping to tap into resentment and distrust from Vietnam and Watergate. But on this topic, Metzenbaum pushed too hard, declaring that Glenn wasn't worthy of the Senate because most of his life had been spent on the public payroll. "How can you run for the Senate when you've never held a job?" he said.

This line of attack proved fatal. To Metzenbaum. It allowed Glenn to harness every element of his beyond-reproach background. The Boy Scout and the warrior, the patriot and the hero responded in a single, devastating voice.

"I served twenty-three years in the United States Marine Corps," Glenn began at a press conference. "I was through two wars. I flew 149 missions. My plane was hit by anti-aircraft fire on twelve different occasions.

"I was in the space program. It wasn't my checkbook, it was my life that was on the line. This was not a nine-to-five job where I took time off to take the daily cash receipts to the bank.

"I ask you to go with me, as I went the other day, to Veterans Hospital and look those men with their mangled bodies in the eye and tell them they didn't hold a job. You go with me to any Gold Star mother, and look her in the eye and tell her that her son did not hold a job. You go with me to the space program, and you go as I have gone to the widows and the orphans of Ed White, Gus Grissom, and Roger Chaffee, and you look those kids in the eye and you tell them that their dad didn't hold a job. You go with me on Memorial Day coming up, and you stand on Arlington National Cemetery — where I have more friends than I like to remember — and you watch those waving flags, and you stand there, and you think about this nation, and you tell me that those people didn't have a job.

"I tell you, Howard Metzenbaum, you should be on your knees every day of your life thanking God that there are some men — some men — who held a job. And they required a dedication to purpose and a love of country and a dedication to duty that was more important than life itself. And their self-sacrifice is what has made this country possible. . . .

"I have held a job, Howard."

Unfortunately for Glenn, other news that day — Jackie Onassis announced her endorsement of Glenn, for instance — overshadowed the fiery speech, keeping it from wide publicity. So Glenn's handlers insisted he give it again, just days later on May 3, 1974, at the biggest campaign event of the season: the City Club debate in Cleveland. Metzenbaum, needless to say, didn't mention the "you've never held a job" issue, so Glenn had to bring it up himself.[10]

"That one hurt, Howard, and it hurt a lot of other people, too," Glenn said at his first opportunity. "It wasn't my checkbook. . . . "

Glenn's old friend Tom Miller still remembers that speech fondly.

"Well, that one thing won the election," he said, "because Metzenbaum had so much money. . . . "

After vanquishing Metzenbaum, Glenn crushed his general election opponent — the mayor of Cleveland — with the largest election total in state history, 1 million votes. Once again, his tenacity had paid off. Two bitter campaign experiences over ten years had failed to deter him from achieving this goal, and his independent doggedness allowed him a rare opportunity in politics. He arrived in Washington having defeated the party machine's candidate, making him beholden to . . . no one.

Even the United Auto Workers, in giving Glenn a $10,000 donation during that general election, sheepishly acknowledged the awkwardness of having to support Glenn after he'd beaten their first choice back in the primary. "We're not giving you this because we backed you in the primary," the UAW official said. "You and everyone here knows we didn't."[11]

But Glenn didn't find it awkward at all. "The nice thing about donations like that is

you don't owe them a thing," he said after getting that union contribution.[12]

Now that he was in office, Glenn set a tone of ideological aloofness that would for many years confound the best efforts to categorize him. He hired a corporate management consultant to organize his Senate office, rather than the typical political advisor, and he scrutinized resumes looking for experience without concern for partisan grounding.[13] He became known rather quickly as a detail-driven, politically indifferent voter who only ventured an opinion on something after he had done lots of homework.

"He was a little less of a political operator," said Michael Barone, a longtime Washington journalist who co-writes the well-respected *Almanac of American Politics*. "A senator who didn't seem to be sympathetic to the political considerations of other senators."

But Glenn was not in Washington long before the national spotlight crossed his bow. And another disappointment was on its way. The 1976 presidential election was coming, and Glenn's name persistently surfaced among the list of potential Democratic contenders, until he — again the novice — forfeited any opportunity to play up the notoriety by declaring in no uncertain terms that he didn't want to run for president. He then became a vice presidential possibility who got a very close look by Jimmy Carter, and Glenn was asked to be one of two keynote speakers at the Democratic National Convention in New York City.

He bombed.

Antsy delegates roamed the floor of Madison Square Garden, ignoring Glenn altogether. Never a great public speaker, Glenn was stiff and bland as he tried to invoke the compassionate memory of Franklin Delano Roosevelt. He talked sincerely about patriotism, and about the Democratic Party's "sins of the warm-hearted" for trying too

While Senator Glenn didn't always vote along party lines, he was still an important ally for President Jimmy Carter.

WHITE HOUSE PHOTO

hard to help too much, contrasting that with what he called the GOP's "sins of the cold-blooded."

Congressional Quarterly, the weekly magazine, in its July 17, 1976, coverage of the convention summarized Glenn's address this way: "Roused momentarily when Glenn was escorted from the Ohio delegation to the podium, the crowd gave Glenn a warm reception when he began to speak. But the former astronaut's low-key presentation and relatively non-controversial approach failed to hold their interest."

Worse yet, Glenn was about to be outdone. Following him on stage was the second keynote speaker, U.S. Rep. Barbara Jordan of Texas. Although she struck many of the same themes in her speech, her delivery electrified the delegation in a way Glenn hadn't even approached. She was the first black person to keynote one of the national party conventions, and she stole the night.

Glenn's vice presidential aspirations vanished with his speech, and he returned to his Senate duties with a new colleague in the chamber. Metzenbaum had unseated Taft, and the two Democrat antagonists began a long, painfully slow thaw of their chilly relationship. Glenn returned to his task of building a record vote by vote that satisfied him, but not always his party. More liberal than his military bearing might suggest, Glenn at times voted all over the map. He supported the Equal Rights Amendment, but often broke ranks and went against organized labor. He was generally supportive of military weapons systems, but opposed the M-X missile that formed the backbone of President Carter's defense program. And all the while, he was developing an expertise in areas of interest to him, among them foreign affairs. But ever the technical thinker, he was doing

so through the details. He became a critical vote to the Carter Administration during Senate consideration of the Strategic Arms Limitation Treaty, or SALT II, with the Soviet Union. But he didn't engage in debate at the philosophical level of the treaty, which Glenn said he wanted to support. Instead he was vexed by a single issue, and he refused to relent. The fall of the Shah of Iran had closed off American access to two missile monitoring sites that would be used to verify Soviet compliance with launch restrictions. Without those sites, Glenn said, he couldn't support an arms limitation agreement that couldn't be verified. He was so steadfast and well-schooled in his position that he actually converted the late Sen. Barry Goldwater, a towering Republican, to share his concern. The Soviet ambassador chastised Glenn for not trusting his country during a Washington lun-

The Democrats wanted to regain the White House in 1984, and among the candidates was John Glenn, who campaigned in his hometown of New Concord, Ohio, with Gov. Richard Celeste.

Glenn campaigned hard in 1984 but didn't fare well in the early Democratic primaries. He hoped for a strong showing on Super Tuesday, but when he didn't get one, he dropped out of the race.

cheon in 1979, but Glenn wasn't swayed.

"If we trusted each other we wouldn't need the treaty in the first place," Glenn said.[14]

Glenn's other big interests were in the plan to reorganize the Postal Service, and in a bill to streamline the employee grievance procedures for Senate staff. Glenn also began burrowing into the gritty details of nuclear technology, developing informed opinions about export controls on this technology and on devising sanctions against countries that began active nuclear programs. Glenn was to become an influential expert in this field, but overall he occupied his time with valuable, though not sexy, government business.

"Well, it would clear out a press table, I can guarantee you that," Glenn said. "But some of the things on efficiencies of government . . . I happen to think that [line of legislation] was one of the most important things we could have taken up. And I took that on."

Glenn's standing in the Senate, perhaps damaged a bit by his convention speech and his bridesmaid finish in the vice president contest, was restored in 1980 with a second huge election victory. While the Carter-Mondale ticket was losing to Ronald Reagan, Glenn smashed out another record-setting win, this time pulling in 1.6 million votes.

And soon, the national spotlight returned. After his reelection, Glenn altered his previous nose-to-the-grindstone approach just enough to begin doing some traveling. Where before he resisted press attention in favor of his homework, now he began making more public appearances, raising his profile. President Reagan was not doing well in the polls — a victim of an economic recession — which emboldened Democratic chal-

ELISE AMENDOLA/AP

Glenn's credentials — senator, war hero, test pilot, and astronaut — made him an exciting Democratic presidential candidate.

lengers. Glenn was among them, and by sheer force of his name he rose to the top of the list of contenders, alongside former Vice President Walter Mondale of Minnesota. But even as Glenn's star appeared to rise, trouble followed. The same old trouble. Can he convince people he sees the big picture? Can he win the Democratic primary? Can he give a decent speech?

In October 1982, *The Dallas Morning News* followed Glenn on a trip around America, where he appeared at campaign stops for Democratic candidates as he tested his viability for a White House bid. In Iowa, a vital early caucus state, attendees of a Glenn speech were less than thrilled.

"He'd be great if he didn't have to talk," the paper quoted an Iowa official as saying. "The speech wasn't bad, but unfortunately he had to give it."

As he did in his Senate votes, Glenn moved with great caution and deliberateness in starting his presidential campaign. Trouble was, he didn't have the luxury of taking too long. Mondale, with many idle days since being tossed from Washington along with President Carter in 1980, had used his free time to organize a well-thought-out national campaign. That Glenn even was considered Mondale's peer in the race for the nomination was a testament to Glenn's likability.

"The fascinating thing about the potential power of Glenn," a political veteran told *Columbus Monthly* for a story in March 1983, "is that he has come so far without the normal attachments. He hasn't had a press operation in his office for more than a year. Where's his political operation, his

ASSOCIATED PRESS

In 1984, seven was important again — the number of candidates seeking the Democratic nod for president. The candidates from left to right: former Gov. Reuben Askew, Sen. Alan Cranston, Sen. John Glenn, Sen. Gary Hart, Sen. Ernest Hollings, former Sen. George McGovern, and former Vice President Walter Mondale.

fund-raising mechanism? He is a formidable presidential candidate despite everything. Is all that other stuff irrelevant? Is the core of his candidacy so strong that it drives itself? Of course he can win the nomination. Look, just by being himself, without the accouterments of candidacy, he's one of the top two."

But such a sanguine, wide-eyed view of Glenn's potential, even by a veteran of presidential politics, gave too much credit to Glenn's personal appeal and not enough credit to the mechanics of being elected to the White House. Even the Hollywood release of "The Right Stuff," a movie based on the Tom Wolfe book glorifying America's original space heroes, among them John Glenn, could not alter political reality. The same *Columbus Monthly* story noted that while polls showed Glenn to be a stronger candidate against Reagan in a national race, Mondale still beat

CHARLES STEINBRUNNER

After his failed presidential bid, Glenn went back to the tasks at hand — paying attention to his Senate duties and meeting his constituents, as he did in Dayton, Ohio, in 1986.

Glenn became known rather quickly as a detail-driven lawmaker who only ventured an opinion on something after he'd done lots of homework.

Glenn badly among Democrats, forty-two to eighteen percent.

And so the campaign went. Just as had happened in Ohio during those early years in politics, Glenn's overall appeal was undercut by his standing within the Democratic Party. He might have been moderate enough to run well against Reagan in the general election, but that made him too moderate to win a nomination to challenge Reagan in the first place. For instance, his valuable work for the cause of nuclear nonproliferation might have made him a legitimate choice for dove-leaning voters in any other time, but Glenn was running in a day when the doves wanted a candidate who supported a no-caveats nuclear freeze.

"He found himself without a party," Barone said of Glenn. "I don't think that's dishonorable."

However, suggestions of dishonor were headed Glenn's way. Although the early primaries made clear that Glenn was out of contention for the Democratic nomination, he insisted on staying in the race through a Super Tuesday primary that covered votes in a batch of states through the South. So many dele-

Glenn led the Ohio delegation in 1996 as it cast the votes that nominated President Bill Clinton as the Democratic Party candidate.

gates were at stake in the single day that Glenn could have made up lost ground on Mondale if he scored some big upsets. But he fared no better on Super Tuesday than he had in previous primaries and was forced to fold his presidential campaign. But that single decision to hang on through Super Tuesday sunk millions of dollars of borrowed cash into Glenn's dying campaign, and it saddled him, for the first time ever, with an issue that questioned his integrity. In getting bank loans to fund his Super Tuesday push, Glenn used "letters of comfort" from powerful supporters, Barone recalled, in which these people promised to help Glenn raise money to repay the loans but stopped short of guaranteeing the money themselves. Barone said this rarely used device was an "end-run around the law" of campaign finance limits that allowed wealthy benefactors to use their influence to secure Glenn more money, without exceeding their contribution limits.

"It was an error in judgment" on Glenn's part, Barone said.

But more troubling for Glenn's future was the fact that he had trouble raising the money to repay the loans anyway. Never a big fan of fundraising, even for his vital Senate seat,

BRIAN BAHR/AP

Glenn gives a thumbs-up during the nomination of President Clinton for a second term.

RICK MCKAY/COX

Glenn admits that his years in politics have not been easy.

Glenn found it increasingly difficult over the years to ask for donations to pay for a dead presidential bid. Ultimately, he won approval from the Federal Elections Commission to use his own money to pay the debts of his campaign, and he did spend a large amount of his personal fortune, but he never paid it all off. Years later, the original bank loans from 1984 were still not paid. The FEC rejected a payment plan in 1997, even though the banks had accepted it. The plan was to repay the principal but no interest on the loans. All this made him vulnerable around election time to charges that he was not the flawless hero that was the central component of his public persona. His closest reelection came in 1992, when Lt. Gov. Mike DeWine, a veteran of Ohio politics and a dogged campaigner, made

the Glenn presidential debts a theme of his campaign. DeWine, who later won the seat Howard Metzenbaum retired, ran television commercials spoofing the popular Energizer Bunny commercial by showing a rabbit in a space suit with an announcer saying, "John Glenn. He keeps owing and owing and owing. . . . "[15]

But the centerpiece of the DeWine campaign was even more damaging. DeWine made an issue of Glenn's involvement in a 1988 meeting with Charles Keating, a wealthy Cincinnati financier whose savings and loan operation was being investigated by federal regulators. Keating had contributed thousands of dollars to Glenn's political campaigns in the past, but hadn't donated anything in the two years leading up to the fate-

BOB DAUGHERTY/AP

Throughout his political career Senator Glenn would use his role in space to his advantage since the image of him as a pioneer in space was a powerful political asset. With the Friendship 7 space capsule serving as a backdrop, he spoke during a news conference at the Smithsonian's Air and Space Museum in Washington, D.C., on February 19, 1987, the twenty-fifth anniversary of his historic orbital flight.

ful meeting. Glenn had been told that Keating's thrift was under criminal investigation, but he "hosted" a lunch between the financier and then-House Speaker Jim Wright. Although it never was established that the bank trouble was discussed at the meeting, the Senate Ethics Committee declared on February 27, 1991, that Glenn "exercised poor judgment" in arranging the lunch.[16] He broke no rules and violated no laws, but being dragged into a scandal was painful to him and to his friends.

"The biggest blow I think that he had was with the Keating situation," Miller said. "Now, I knew pretty much what happened in that, and you know, he didn't do a damn thing wrong. . . . Glenn was always pretty upset that he was carried all the way through it, and it cost him a hell of a lot of money in lawyer fees defending himself. It hurt him because it questioned his integrity. . . . "

Still, Glenn survived the DeWine scare and maintained his position among the Senate's most famous and most respected members. In the summer of 1997, when the Senate Governmental Affairs Committee was assigned the task of investigating allegations that campaign finance laws were broken during the 1996 presidential race, both Republicans and Democrats were openly pleased to have Glenn leading the Democratic side of the committee because of his reputation for nonpartisanship. Glenn did, however, fiercely attack the Republican-led inquiry, charging it with an unfair focus on the Clinton-Gore campaign. The Senate hearings on the topic produced little to confirm abuse of laws, but it did create an opportunity to question Glenn's motives when NASA announced six months later that the soon-to-be-seventy-seven-year-old senator was cleared for a space shuttle mission. The suggestion by critics, many of them anonymous, was that Glenn's shuttle trip was a payoff for the way he protected the White House during the campaign finance probe.

But by now, Glenn had learned a lot about politics. He bristled at those allegations, emphatic that he never once spoke with the White House during the campaign finance hearings and quick to explain that his idea for a shuttle mission was in the works long before those hearings, but he never lost his cool.

His years in political life were not easy. In fact, they came at a sometimes significant personal expense.

"Absolutely," Glenn said. "So what do you do? What do you do then? Do you say, 'OK, stay away from it?' That's the death knell of a real democracy, if you tell people to stay away from participation in government and politics. . . . If I can just impart a little bit of the feeling I have about the importance of politics, and the importance of government in general, to some of the young people up there and talk to them about it, then, well, maybe that's a good purpose for us."

But first, again, space.

NOTES

1. Text of Col. John Glenn's statement as printed in *The New York Times*, March 31, 1964.

2. "Glenn Got Rich from Ground Up," *Dayton Daily News*, Andrew Alexander, December 7, 1983.

3. *Dayton Daily News*, Alexander.

4. *Dayton Daily News*, Alexander.

5. "Glenn Sets Out on His First Campaign Trail," *Dayton Daily News*, Tom Price, December 8, 1983.

6. *Ohio Politics*, Edited by Alexander P. Lamis (The Kent State University Press: Kent, Ohio, 1994), p. 95.

7. *Dayton Daily News*, Price.

8. *The Almanac of American Politics*, Michael Barone and Grant Ujifusa (Barone & Co., Washington, D.C., 1982), p. 848.

9. *Dayton Daily News*, Price.

10. Lamis, p. 211.

11. "Success Let Glenn Do It His Way," *Dayton Daily News*, Tom Price, December 9, 1983.

12. *Dayton Daily News*, Price.

13. *Dayton Daily News*, Price.

14. *The (Cleveland) Plain Dealer*, George P. Rasanen, November 25, 1979.

15. *The Washington Post*, Ken Ringle, October 22, 1992.

16. Text of Senate Ethics Committee report of hearings on the Keating Five issue, as printed in *Congressional Quarterly*, March 2, 1991, p. 565.

CHAPTER 8
THE JOURNEY

Houston in April 1998. The Lyndon Johnson Space Center. A carnival-sized crowd in blue suits and neckties abruptly pulled stakes at the end of the speech, and the carpeted warehouse in which NASA keeps its biggest space training equipment ebbed into a hush.

The departure of the President of the United States can do that to a room, even one this big and industrial. It can drain a place of so many people so quickly it feels like the air has left, too. Spectators, security officers, TV cameras with their clacking bird legs, all suddenly gone. Out.

But not everyone left with President Clinton when he ended his tour of the space center that day. There among the hulking hulls of shuttle trainers, the seven Discovery crew members kicked around on NASA's royal blue carpet, still buzzing a bit from the VIP glow. Still excited about the face time with the president.

With one exception.

Payload specialist John Glenn was not high on the thrill of having been with the president; he did that often enough in his day job. In fact, he wasn't there to bask in Clinton's aura at all. If anything, just the opposite seemed more likely. There was little doubt that Clinton, dogged by sex scandal, could use the wholesome imagery on the evening news of him shoulder to shoulder with a True American Hero. And Glenn, while he was eager to tell the president all about his mission, was not sorry when Clinton finished the tour and headed on because it opened up a rare moment priceless to Glenn himself: a chance to hang out with his spaceship crew.

Away from the glare, off camera, Glenn reveled in the camaraderie of being John rather than Senator Glenn, dribbling out M&Ms to all hands, quietly swapping tastes of the leftover space food Clinton had ceremoniously sampled before a battery of cameras and lights. Glenn laughed in the glory of it, biting a steaming fork of rehydrated shrimp cocktail offered by a crewmate who declared it "the Cadillac of space food."

"Oh, this is great!" Glenn said, mouth open, chewing, hot. "May have shrimp for breakfast, lunch and dinner, I don't know. I found out one thing in tasting it: stir it up a little bit. If you get down to the corner (of the plastic bag) where that horseradish is concentrated, Ohhhhhh!"

Even the fire of the horseradish was good. Soon enough, he would be doing this in outer space, back on the celestial trail he had blazed half a lifetime ago.

America had months to wait for Glenn's return to glory with Discovery's launch in

During the April 1998 visit, President Clinton — flanked by the Discovery shuttle crew — said it's important that people in their mid-seventies be held as role models.

October, but for Glenn the glory had already begun. Here he was learning how to use the myriad systems of a space vehicle exponentially bigger and more sophisticated than his tiny Friendship 7 capsule. Here he was, poring over experiment manuals, dangling down the side of a life-sized shuttle trainer, and mastering — at long last, since he was so slow warming to the new computers in his Senate office — the high-powered laptops that run some of the science experiments aboard Discovery.

"I'm enjoying every minute of it," Glenn said that day in Houston. "I want to spend all the time here I can getting ready."

Believe that.

Any man in his seventies who talked openly about a desire to fly in space would be considered naive, if not a fool. Even a retired astronaut would know better than to suggest that NASA give him another ride into orbit ahead of any of the scores of young, strong, highly trained men and women in the astronaut corps.

Anybody but Glenn. Who else but Glenn could stand in front of newspaper reporters and say with a straight face that he would like to go back into space again?

Who else but Glenn could do it without being laughed at?

Who else but Glenn could take his biggest obstacle to another spaceflight — old age — and turn it into the very reason why NASA should send him?

Glenn claimed it was all for the sake of

President Clinton said he was heartsick that Glenn was going to retire from the Senate, and made jokes about his return to space during the April 1998 visit to Johnson Space Center.

science. He said he was struck by the similarities between the effects of spaceflight and the aging process. He thought NASA should send an old person into space to study the similarities, and he volunteered himself as the ideal candidate.

But the odds are good that had anybody else taken that same proposal to NASA, he would have been ignored at best or, more likely, laughed out of the office.

But nobody laughed at Glenn. Not NASA Administrator Daniel Goldin. Not the press. Not the public. Oh, there were the jokes. Retired astronaut Jim Lovell, who command-

ed Apollo 13's perilous flight around the moon in 1970, didn't try to conceal his envy of Glenn's new mission. "I offered to be his backup, but they said I was too young," quipped Lovell, Glenn's junior by six years.

And there were critics. Even astronauts who thought the science of Glenn's age study was sound questioned Glenn's selection over other elder astronauts with much more space experience. Glenn had spent less than five hours in space, while many later astronauts — including some of his Mercury 7 colleagues — had logged days of space time. And there were several aging astronauts

NASA

President Clinton's visit included a tour by Glenn.

with space shuttle experience.

It was a point retired astronaut Story Musgrave made bluntly after Glenn held a press conference at Kennedy Space Center on January 21, 1998, just five days after NASA chief Goldin had announced Glenn's flight.

Musgrave, a former Marine Corps aviator like Glenn, held the title of oldest astronaut at age sixty-one when he made his sixth and last spaceflight on Columbia in 1996, already a year older than the mandatory retirement age for U.S. airline pilots. Musgrave was still flat-bellied and muscular as he sat in the press

center conference room surrounded by a small group of reporters who had hung around after Glenn strode out, taking most of the crowd with him. He was there as a consultant for CNN, but he became the subject of an impromptu press conference when several reporters asked what he thought of Glenn's mission.

Although Musgrave said he had retired after NASA told him his sixth shuttle flight was his last, he insisted he wasn't envious of Glenn, and he wasn't going to seek reinstatement in the wake of Glenn's selection. "I've had the privilege of thirty years in the astronaut corps," he said.

But he questioned how much Glenn could contribute to the mission with no prior shuttle experience and only limited training time in the coming months. Astronauts typically spend a year training full-time for a mission, Musgrave said; Glenn would have to squeeze his training into breaks in his Senate schedule between January and October.

Musgrave said he was disappointed when he learned Glenn would not train full-time for the mission. "Like anything else, you get out of it what you put in," he said. Besides limiting his own work, Musgrave speculated, Glenn's lack of familiarity with the shuttle's complex systems would mean other crew members could find themselves helping him out with a lot of minor tasks, cutting into their own time.

Musgrave was all for giving Glenn a shuttle flight, but not on any pretext of science. "NASA should send Glenn [into space] because of who he is," he said.

NASA agreed. But not right away.

It was an inescapably nostalgic notion. Good old Glenn, wanting one more fiery ride to glory. There was an attractive symmetry to the idea. The hero of the space program when it was a weapon of the Cold War, returning to the space program now that it's an instrument of international cooperation. Never mind the science: It would be great PR for NASA to hurl its most prized hero back into the heavens.

But NASA had tried giving shuttle rides for PR before, and the scheme had blown up in its face.

In 1984, President Reagan announced the Teacher in Space program, a plan to train an ordinary schoolteacher for a space shuttle mission. It just so happened that Reagan was running for re-election against Walter Mondale, and Mondale had just won the endorsement of the National Education Association. NASA

selected Christa McCauliffe to be its first school teacher in orbit. But it didn't stop there. Putting a teacher in space would just be the first step in what the agency called its "Spaceflight Participation Program" — an initiative to open the space frontier to citizens in nonspace fields. It started taking applications from journalists. And it gave space shuttle rides to U.S. Sen. Jake Garn, R-Utah, chairman of the Senate Appropriations Committee, and U.S. Rep. Bill Nelson, D-Fla., whose district included Kennedy Space Center.

Glenn wasn't about to oppose his colleagues' junkets. But he drew the line at teachers and reporters.

"It's more of a political ploy than anything else," Glenn said in December 1985 of McCauliffe's scheduled flight. "I'm not against other people going into space," he said, but added, "The basic purpose of the space program is basic, fundamental research, and I think that's what we should be concentrating on. The public support for the program is going to come as we discover new things, new materials, new benefits, and as the public understands the space program is bringing these benefits, and that we're not a bunch of people joyriding around. What I have not wanted to see is every butcher, baker and candlestick maker thinking he has a right to get on."

Because it can be dangerous. McCauliffe's shuttle flight lasted seventy-three seconds. A failed joint on one of the shuttle's twin solid-fuel booster rockets spurted flame and ignited the shuttle's massive propellent tank. A spectacular explosion tore the Challenger orbiter to pieces. Its crew cabin, with six astronauts and a schoolteacher inside, plummeted like a rock, down through the sky for more than eight miles, into the ocean, while parents, teachers and schoolchildren all over America watched in horror.

NASA and the space shuttle program sur-

Glenn met with President Clinton for a close-up look at preparations for the STS-90 shuttle launch. On either side of Glenn are NASA Administrator Goldin (left) and JSC Director George W.S. Abbey.

vived, but NASA officials wanted nothing more to do with private citizens in space.

To this day, Glenn insists his whole space shuttle thing began accidentally. He swears he wasn't fishing for a way back into the astronaut program, to get that second space shot he was denied during Project Mercury. Glenn was preparing for NASA's budget hearings in the summer of 1996 when he stumbled across a medical chart that listed the multitude of physiological reactions the human body has to prolonged weightlessness.

He saw muscle atrophy and cardiovascular shifts, and sleeplessness and weakening of the immune system. Brittling of the bones. Dozens of changes to the body that astronauts endure in weightlessness before recovering fully soon after returning to the steady pull of Earth's gravity.

Glenn had a thought. He snatched a medical book on geriatrics, and there he saw another chart listing the many painful effects of normal human aging.

As a man in his seventies, Glenn was becoming familiar with some of these conditions, though he had been stalling them as best he could with a diet of perpetual moderation and a devotion to brisk two-mile walks four or five times a week. NASA already was working with the National Institute of Aging, a part of the National Institutes of Health, on some of these same issues, but experts there hadn't yet proposed sending an elderly person into space to see if an aged body would react any differently than younger astronaut bodies do.

"I got into this thing and developed what I thought was a rationale for somebody [in my age group] going to look into these particular areas," Glenn said. "And if I could do it, why, fine."

But Glenn admits it wasn't really that casual for him. What he saw in those medical books was more than just a few interesting facts that made him curious. What Glenn saw in those medical books was a mission.

"I really did go in and say somebody ought to do this. Whether it was me or not," Glenn said. "But, right along with it was, 'And look, Dan, I'd like to be the guy that does this.' There wasn't any doubt about that."

Within months he was on the floor of the Senate reading from NASA documents about the similar effects of aging and space travel, sharpening his rationale, taking it to the public through C-SPAN.

"I wish I could have the very personal attention of every person in this country who is sixty years of age or older," Glenn said from the floor. "One thing that has happened in the look into the life and biosciences in the NASA program has been that we find some notable parallels between what happens to astronauts in space and what happens to the elderly right here on Earth.

"And if we can find what triggers some of these similarities, perhaps we will have a whole new handle on approaching difficulties that people have right here on Earth."

Goldin, a President Bush appointee, suddenly found himself being asked to approve sending into space a man born not many years after the Wright brothers triumphed at Kitty Hawk. And of course, the issue was more than age because this was not just any fit-as-a-fiddle grandfather. John Glenn is a national icon, a Cold War and Real War hero, the very personification of America's pioneering work in space.

In an interview five months after he made the decision, Goldin acknowledged the symbolic aspects of flying Glenn weren't lost on him. Nor was the fact that Glenn had paid a price for becoming a national hero — the highest price an astronaut could pay. President Kennedy had decided Glenn was too great a treasure to risk on a rocket, and Glenn had retired while his Mercury buddies went on to fly higher, faster, and farther. Al Shepard went all the way to the moon.

"I acknowledge up front that science wasn't . . . the only reason," Goldin said. "I felt he deserved a second flight, but he had to do cutting edge, peer-review science. That was mandatory."

Still, nobody wants to be the guy whose decision to send America's John Glenn back into space accidentally kills him. And flying Glenn would mean crossing into a new medical frontier.

"We were in uncharted territory," Goldin said. Glenn would be sixteen years older than the oldest astronaut who had ever flown. He would be well into that stage of life where aging has begun taking its toll. The very changes that scientists wanted to study were those that raised concerns about Glenn's ability to withstand the rigors of spaceflight.

Goldin made a list of requirements that

would have to be met before he would agree to the flight. The science experiments would have to be legitimate, and Glenn's health would have to be a certainty. "I wanted the *i*'s dotted and the *t*'s crossed," Goldin said. "And it literally took about a year to do that. . . . There was a very thorough set of tests, and some of them were quite intrusive, but because of medical privacy I cannot discuss them."

Glenn said he told his fifty-two-year-old son David, a physician in California, about all the medical tests he'd undergone, and the younger Glenn concluded that NASA had done everything but an autopsy.

Politics was another issue. The NASA administrator serves at the pleasure of the president, and Goldin said he wanted to be able to make his decision without political pressure. He said he made Glenn promise not to recruit White House support, and he said no pressure ever came from there. "The president never called me, the vice president never called me," he said. He said his only contact with Clinton and Vice President Al Gore regarding Glenn was to notify them of his decision.

But there was someone whose approval Goldin did need: Glenn's wife, Annie.

"I know her and I called her," Goldin said. "You know what she said to me? 'This is what John wants to do. I've been with him for fifty-five years, I'm going to support him.'"

Selecting Glenn for a flight was really a decision Goldin didn't think he would have to make. "First, I didn't think he was going to pass the physicals. . . . I knew it was going to be tough. And second, I thought it would be awfully tough to get some good peer reviewed science. I'm a very tough taskmaster," he said.

Goldin underestimated the MiG Mad Marine.

Glenn, after all, served on the Senate Select Committee on Aging. He dug into the research. Goldin is still incredulous at the memory of Glenn lumbering into his office with two armloads of scientific literature for their first meeting on the topic. Unable to describe it without demonstrating, the NASA chief hopped up from his chair and bent over like an ape, arms hooked as if loaded down with books and papers. "I mean, he walked into this office like this," Goldin blurted, "with books, and research that he did. . . . It was awesome."

But Goldin said he wanted more. "I said, 'You're going to have to get the scientific community behind you.' So he said, 'OK.' And then he went out and he worked with the National Institutes on Aging, and they called a workshop on this whole subject. And I think 50 or 100 people showed up. . . . I mean this guy, he took my orders very seriously. He went out and did it. And one of the recommendations was that we ought to consider flying between five and ten people between [the ages of] seventy-five and eighty."

Ultimately, Goldin met with the directors of the National Institute on Aging and its parent organization, the National Institutes of Health. Without Glenn. "I didn't want any intimidation," Goldin said. "I talked to them about it because I wanted to be sure that I could tell the American public that this is good science."

Meanwhile, Glenn was in the zone. Whatever Goldin wanted, Glenn was willing to do twice as much. He started calling Goldin's staff. Just to check. How's it going? What's cooking on my idea for a shuttle flight? Have you heard anything new? Can I get you any more information?

Glenn insists he wasn't trying to pressure Goldin. "He had designated a couple of people to handle the contacts on this and keep me apprised of what they were doing, and I'd keep them apprised of what I was doing,"

Glenn said. "I'd be sitting here [in his office] someday and I'd think about it, and I'd pick up the phone and call over there [and ask] 'Hey, what have you heard, what's cooking?' But there wasn't any, you know, I had no planned program of so many calls or anything like that."

But Glenn's persistence was driving Goldin's people nuts. "He must have called . . . fifty times," Goldin recalled with a laugh. "This is his intensity. Oh, he pushed. He pushed."

Before long, the phone rang in the Arlington, Virginia, home of Glenn's good friend Tom Miller. This was the same Arlington house Miller lived in back in 1962, which was next door to the house the Glenns owned at the time — the Millers and Glenns built the houses side by side — the house where Annie Glenn and the two teenage Glenn kids watched Glenn's first spaceflight, and where Vice President Lyndon Johnson dropped by that historic day.

The voice on the phone was Goldin's. He needed suggestions for dealing with Glenn, Miller said. All those phone calls. How do I slow him down? Goldin wanted to know. But Miller didn't have much advice to offer on taming this guy he'd known since flight school fifty years ago. It was more like a warning.

"I told Dan, 'You're in contact with the most persistent person I've ever known,'" Miller said.

On Thursday, January 15, 1998, Goldin closed the door to his office, sat down at his desk, and told himself it was time to decide about Glenn. "I threw everyone out of my room. And I sat in here, and I scripted all the issues. I had a little checklist that I made on a scrap of paper, and I went down the checklist to satisfy myself that everything was done right. . . . It was a very big decision that had to be made. You know, it was not a comfortable decision," he said.

But everything added up. Glenn had passed all the medical tests. The science made sense. The White House had kept its nose out of it. And Annie supported it. Goldin had his staff notify the White House, then called Glenn out of a meeting to give him the word.

Meanwhile, word was leaking to the press. Phones started ringing. Aides were itching to tell the world that Glenn had his shuttle flight, but no one would go on the record; this was Goldin's announcement to make. Glenn's office issued a terse statement that hinted at everything but confirmed nothing: "I understand there is a great deal of interest in this matter, but today I have no comment on it. I look forward to discussing this in the future." Goldin's office issued a media advisory that he would make a "major announcement" at noon the next day, but there was little doubt what it would be. Glenn's flight was a national story the evening before the news conference.

So Goldin could only confirm what the world already knew, but he held a news conference at NASA headquarters that included Glenn and the heads of the NIA and NIH. He wanted to make it clear that this wasn't a senator's junket or the last trail ride of an American cowboy, but good, hard science.

But the symbolism was impossible to ignore. "What a great day for America," Goldin gushed. "The man who almost thirty-six years ago climbed into Friendship 7 and showed boundless promise for a new generation is now poised to show the world that senior citizens have the 'Right Stuff.'"

Indeed, outside medical circles, few took science as more than an excuse for Glenn's flight. It was either the ultimate salute to an American hero or the ultimate junket. Editorial cartoonists had a field day with the aging issue. There was Glenn, swimming in zero-g after his dentures. There he was again, being reminded by Mission Control that he'd

Opening the Space
Frontier

Goals
• Explore and settle the solar system
• Achieve routine space travel
• Increase human knowledge
  of nature's processes using
  the space environment
• Enrich life on Earth through people
  living and working in space

...SA Strategic Plan

RICK MCKAY

Glenn — with NASA Administrator Daniel Goldin — was the first American in orbit and would now be the oldest American in space.

left the shuttle's turn blinker on.

Cynics saw the Glenn launch as a good start: "I'm in favor of sending Glenn up into space. I'm just vehemently opposed to bringing him back. Send the whole Congress up there and keep them up there," said a reader who phoned the "Speak Up!" column of one of Glenn's home-state papers, the *Dayton Daily News*.

Space advocates hoped NASA might finally be loosening up its post-Challenger policy against putting ordinary citizens in space. Simultaneous with Glenn's announcement, the agency said it would start training school-teacher Barbara Morgan for a shuttle flight. Morgan had trained with McCauliffe as her backup, and after the Challenger disaster

NASA said she would be the first choice for the next teacher in spaceflight. But since NASA had no stomach for that idea, being the first choice for a civilian in space program simply left her in limbo for a dozen years. Now, space advocates thought the paired announcements were more than a coincidence. "While this falls short of reopening the civilian space program," said the National Space Society in Washington, D.C., the announcements "suggest that flight opportunities will be available to people other than career astronauts in the future."

A generation after the flight of Friendship 7, many observers saw Glenn more as an ordinary citizen than a returning astronaut. But NASA denied any relationship between the two flights. This was not a resumption of the Spaceflight Participation Program, NASA spokeswoman Debra Rahn said. "You should not put the two together," she admonished. "Sen. Glenn would not be considered a civilian."

After the press conference, NASA held an informal coffee-and-doughnuts reception. Employees started wandering in. Not for doughnuts, but for pictures. Of themselves. With Glenn.

"Now these are hardened individuals who have been through the scene," Goldin said in a tone of wonderment. "There must have been about two or three dozen, at least, maybe even more, four or five dozen people. They kept meandering up, shuffling their feet and saying, 'Sen. Glenn, can we get a picture with you? Could you give me your signature?'"

The simple, inescapable fact was that many Americans supported Glenn's flight more for its inspiration than for any science that might come from it. But Glenn would never acknowledge that as a reason for going.

"Whether it will be an inspiration, I don't know. . . . Will it make me an inspiration to older people to be more active or will it affect younger peoples' view that the elderly are not people just to be put away in a bin? You know, will there be that aspect to it? I suppose there will. But that wasn't the design when we were starting out. We didn't set out to design it to do that," Glenn said.

To be fair, at least some of Glenn's reticence may have been to deflect the charges of critics who say his flight had little scientific value and was instead a nostalgic adventure. But it was frustrating to people like John Pike, director of space policy for the Federation of American Scientists, who was convinced Glenn's flight was all about symbolism and inspiration and very little about science.

"The question you have to ask is, 'Why are we doing this?'" Pike said. "We're doing it because it's John Glenn; we're not doing it because he's a geezer. . . . I wish he would get off this medical stuff and give us some pep talks."

Although Annie wasn't steeling herself against the criticism, she was a bit like her husband when it came to talking about his new mission. More than likely, she was protecting herself against the deluge of interview requests that stacked up throughout her husband's training, and perhaps also against her own apprehension about what he was doing. His true love since high school, she had been stiff-upper-lipping it for a long, long time. In July, milling about at Johnson Space Center while Glenn and the crew worked through the routine they would follow upon first achieving orbit, Annie shrugged at questions about the dangers involved.

"It's not the scary type like it was before because back then it was so, so unknown," she said. "I had confidence then, but you just didn't know. But he was a test pilot, so that's just one of those things."

At a suggestion that he still was, she laughed and nodded approvingly. "He keeps me going. He keeps me alert."

Although the Glenns concede that the family was at first "cool" to the idea of his return to space, everyone eventually climbed aboard. Glenn had both his children and the two grandkids out to Houston for a tour of the simulators. And Annie was with him for most of his sessions at the space center. She said seeing things firsthand helped her overcome the butterflies and even catch some of his enthusiasm.

"I'm very lucky that I can learn what's

In 1996, John Glenn began thinking about going back in space, with a scientific purpose in mind. NASA finally agreed and announced on January 16, 1998, that Glenn would see the heavens again, this time from a space shuttle.

going to affect him, and I can watch him in training or attend classes. And he's really so excited. So tickled," she said.

But for the science, Glenn emphasized in every interview. The science. After every question about what the flight would mean to him personally, Glenn would swing back to the mission like a compass finding true north. "I always wanted to go up again," he would say, "but that's sort of beside the point. . . . "

Until finally, one day in his Washington office with the Senate computer behind him cued to the shuttle Discovery's Web page, he let a door drift open.

Yes, there was something about the first spaceflight that moved him. And held him. In a rare, fleeting moment, he confessed that much. Behind the Cold Warrior exulting in the power of American ingenuity there was this guy from Ohio who as a kid made model airplanes and as a grown-up got to look down on the Earth the way angels do. He agreed, then, that it was spiritual.

"Not spiritual in the idea that you expect to run into God or anything like that," he said. "But a spiritual quality in that having such a different vantage point to look at creation, even though you're not that far away from Earth, but to look at it from that new kind of vantage point can't help but give you a greater appreciation for all this creation we live in."

But just that little admission had him squirming. This was clearly off the mission. Tenacity and discipline, the hallmarks of his life, don't help explain the way a single day in orbit — four hours and fifty-five minutes in a metal can no bigger than a refrigerator — could grip a person's imagination for the rest of his life.

"I suppose when you've had an experience like that you're a little different person when you come back," he said. "You've seen things that very few people have ever seen, and that can't help but . . . I guess it maybe makes you . . . maybe a little more tolerant of life in general or you're not quite so prone to be critical of everything that happens."

He shrugged.

"Maybe that's a poor way to put it, I don't know."

Even though Glenn tried to focus on the science of the mission, the nostalgia kept creeping to the surface.

CHAPTER 9
THE MISSION

Dangling by parachute straps over an indoor swimming pool, clad in a bulky orange astronaut suit that seemed to be consuming him, the senior U.S. senator looked to be suspended not just from the ceiling, but from his senses as well. The man is in his seventies, after all. Why would he want to do this? But there was John Glenn, slung over the water like a hapless local celebrity in the charity festival dunking machine, seconds from the moment when somebody drills the bull's-eye. His face, barely visible inside the bulbous helmet, tensed with stern anticipation.

Abruptly, the straps let go. Glenn plunged into the water with a deep KAWOOOSH! The inflated bladders of his life preserver pulled him instantly back to the surface as a life raft tied to his harness exploded into shape behind him. Fighting the bulky suit, the heavy boots, the gloves, the helmet, and the water, Glenn found the raft

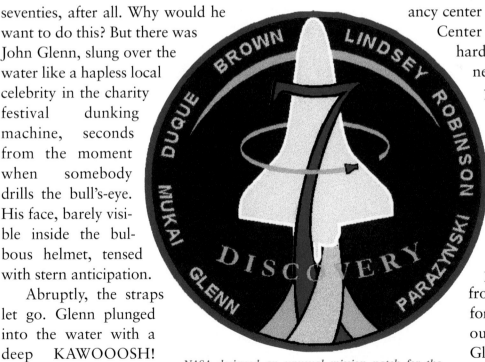

NASA designed an unusual mission patch for the Discovery flight. The large '7' signified the number of crew members, while also recalling Glenn's Mercury flight.

and, with much thrashing and splashing, wriggled his way onto it.

"Well, you're in that suit, and it's a lot of work," Glenn said later. "I wound up sleeping good that night, I'll guarantee you."

The exercises in NASA's neutral buoyancy center at the Johnson Space Center in Houston were hardly easy things for a nearly seventy-seven-year-old man to do. And they had nothing to do with Glenn's scientific mission: If all went well, the flight would be a breeze compared to this. But if some calamity prevented Discovery from reaching orbit and forced its crew to bail out over the ocean, Glenn would have to be able to take care of himself just like every other crew member. Survival training is hard work for people half Glenn's age, but he soldiered through it, grim-faced and unblinking. He had asked for one more spaceflight—pushed with all his might to

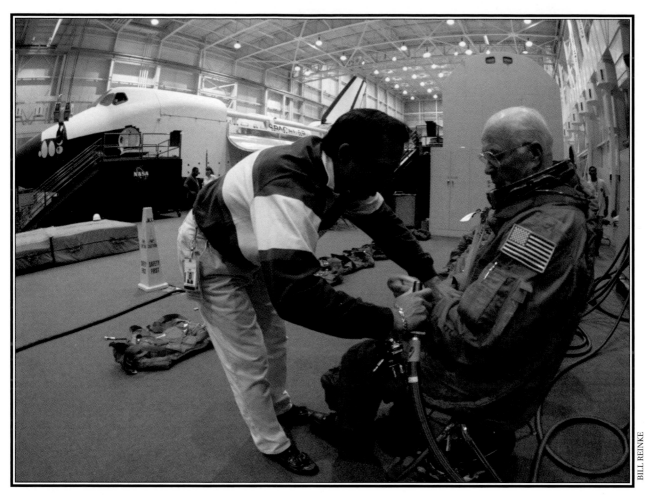

BILL REINKE

Just like all the other astronauts, getting suited up took time . . .

get it, as a matter of fact — and if this was the admission fee, he was only too happy to pay it. And if the fee were even higher — a possibility the survival training itself underscored — he was willing to pay that, too.

For Glenn, this was one more chance to explore the frontier — not of space this time, but of science.

In 1962, Glenn went into space as a test pilot. "We were trying to find answers to some very, very fundamental things," he said. What was it like to try to control a vehicle in space? How should the cockpit be designed? Science questions were just as basic, Glenn recalled. "Such things as, does your eyeball change shape when you're weightless?" One of his Mercury experiments was reading a miniature eye chart every twenty minutes to

see if his vision changed.

"We're so far beyond that now," he said. "We're working in space. We're not just up there getting impressions. We're doing research."

That change, from space exploration to space operations, also recast Glenn from test pilot to researcher. Instead of one man against the cosmos, he would be one in a crew of seven, and the lowest ranking at that. The mission commander and pilot are the ones who actually fly the shuttle. And three others aboard Glenn's flight were named mission specialists, full-time astronauts trained to operate shuttle systems as well as do experiments. But payload specialists like Glenn come from other professions, and they get limited training for specific missions. Glenn

. . . and help.

would be one of two payload specialists on the mission. The other, Dr. Chiaki Mukai, was a Japanese cardiovascular surgeon who had logged 354 hours on Columbia in 1994. Next to her, Glenn was almost a rookie with not quite five hours of spaceflight thirty-six years ago. And while Curt Brown and Steve Lindsey guided Discovery through the heavens, Glenn's top mission priorities would be down in the windowless middeck or science module, logging blood and urine samples and wiring himself for sleep studies.

But if Glenn's status made any difference to him, he never showed it. He saw his return to space as a chance to push back the boundaries of the unknown, which made this low-grade research position as much of a frontier assignment as his hero turn a generation ago.

"I really believe that what I'm doing on this [mission] will be the toe in the door that opens up a whole new area of research for NASA," he said.

All he had to do was kick up into that life raft, learn how to operate a warehouse full of fancy camera equipment, master computers, learn to use a vacuum-sealed toilet, study graduate-level science manuals, and maintain a respectable voting record in the United States Senate. Glenn was putting his toe in the door, all right, but which door? How many? How could he get all of his training done without letting his Senate work slip? Could he handle the rigors of survival training?

Glenn's answer was a firm yes. He couldn't imagine not handling it. Any of it. He wouldn't skip any of the training, even if it were possible,

BILL REINKE

The bulky, heavy orange space suit was much different from the one Glenn wore thirty-six years before.

and as far as NASA was concerned it wasn't possible. "I would not expect anybody to give any waivers on something like that," Glenn said. "I'll carry my full load on that and train for it just like anybody else." So every time the Senate let out, Glenn jumped in his plane and dashed to Houston for an intensive week or two of training. If he had only a long weekend open, he left his six-seater at home and flew commercial to Houston because it was quicker.

The days in Houston were long.

"When the senator's down here we try to keep him busy because we do have him for a limited period of time, and we want him to be as prepared with all of the orbiter systems as we can," flight commander Brown said during a training session break. "Plus, he has a whole suite of life science experiments. . . ."

Then Glenn stepped in: "Those take a lot of time to get up to speed on, but not as much time as Curt and the crew [need] to fly this thing."

But when Glenn was asked what he found

BILL REINKE

Glenn walks toward the centrifuge at Brooks A.F.B. for his ride. His body was given three Gs of gravitational force in his two nine-minute stints.

toughest about the training, Brown answered for him: "I think [it's] just the sheer number of hours we're putting him through while he's in town. We're keeping him very busy, and trying to get as much crammed in as the time allows."

Survival training doesn't take an Olympic athlete, but trainers say it isn't a walk in the park. Just wearing the Advanced Crew Escape Suit takes work. The "pumpkin suit" isn't designed for space walks, but rather to protect an astronaut if a space shuttle cabin suddenly loses pressure during launch or reentry. It isn't too heavy — about twenty pounds — but it comes with a harness weighted down with survival gear: life preserver, life raft, oxygen bottle, drinking water, and flares. All that hardware adds about another twenty-five pounds. (The parachute adds yet another twenty-seven pounds, but astronauts only wear it when

they're strapped into their seats.) Glenn had to waddle through his training sessions with about fifty pounds of gear hanging on him. "It's a load. You have to be in good shape," said Stephanie Walker, manager of crew escape equipment for the shuttle program.

Physical training started in February 1998 at Brooks Air Force Base in San Antonio, Texas. There, clad in his orange suit and strapped into a capsule on the end of a long arm, Glenn was whirled around and around by a 1,000-horsepower motor. The machine, a centrifuge, simulated the heavy acceleration that astronauts would feel during launch and reentry. Acceleration is measured in Gs, with each G being equal to the normal force of the Earth's gravity. Glenn's centrifuge training included two nine-minute rides that peaked at 3 Gs. A cakewalk for a guy with Glenn's background. He had experienced more than

BILL REINKE

Training sessions included learning emergency evacuation procedures, including an escape through the shuttle's top hatch that required astronauts to practice repelling down the side of the orbiter.

he told an operator who was watching him on a TV screen in the control room. "It's like lifting dumbbells without having any in your hands."

Glenn — and the operators — got a nasty surprise at the end of the first run when the centrifuge made a hard stop to simulate the sudden end of acceleration when the space shuttle's main engines stop firing. Glenn's seat had come loose, and he rolled harshly as the arm jerked to a stop. But Glenn took the upset in stride. When the session was over, he even claimed he enjoyed it. "Three Gs squash you down, but you come right back from it," he said.

June found Glenn learning various emergency methods for getting out of the orbiter after landing. One was an inflatable slide that popped out of the side hatch. Another required him to climb up through an overhead window that doubled as an escape hatch and use a rope to rappel down the side of the orbiter. All in that heavy suit. One day, NASA allowed the press to watch and take photos. Glenn talked briefly to reporters afterward, but his eyes had the empty look of true fatigue.

"That is a lot of work for anybody, no matter which crew member it is," trainer Sharon Jones said later. But, she added, "John handles it quite well."

Indeed, Glenn seemed to relish the training — especially the camaraderie, the teamwork, the clarity of purpose. "It's such a pleasure to be working with a bunch of people . . . that are so bright, and are so channeled in one direction," he said. "You've got an objective that everybody's working toward. So different from the Senate where, you know, we've had a lot of rancor creep in around here the last couple of years. . . . [In training,] you've got a purpose and you know what it is and everybody's

twice that much when his Atlas booster punched him into orbit and his capsule slammed back through the atmosphere during reentry. But that was nearly half a lifetime ago for him.

"We use a little bit more finesse coming back into the atmosphere and going out of the atmosphere," Brown said of the shuttle. "So, it's a little bit gentler ride."

Glenn endured the session cheerfully. When the centrifuge reached 2.5 Gs, he raised his arms above his head, slowly waving them over his face. "Just checking arm movement," working together. And they work very, very hard. Very hard."

Glenn claimed his toughest challenges come from the small things — the shuttle's laptop computers, for example. And Glenn was assigned to do much of the inflight photography, which put him into hours of instruction on the use of assorted still and video camera equipment. He'd photograph the crew at work inside the shuttle, as well as capture the views outside. So intensive was the camera study that Glenn would emerge from his space shuttle training nearly qualified to get a job as a television news cameraman. This in addition to the stacks of books and documents he had to study to learn the shuttle's myriad systems and the science he was expected to tackle during the flight. At Johnson Space Center, when they weren't working out in the big training rooms, they were hitting the books in a classroom-like crew office with windows that overlooked the sprawling, flat campus of the space center.

BILL REINKE

Glenn's training was physically demanding, but he did not ask out of any of it.

"We were just test pilots," Glenn said of his Mercury years. "I mean, you had to be smart, but now, good Lord . . . you're bent on some of the highest levels of scientific research going on in the world."

Since Glenn's '62 flight, all kinds of spaceships and space stations have been rocketed into orbit, many of them dwarfing his tiny Friendship 7. The entire fleet of Mercury capsules flown in the '60s would easily fit inside Discovery's payload bay. But for all the advances in size of spaceships and duration of spaceflights, surprisingly little medical research has been done. Sure, NASA has drawerloads of vital stats on every astronaut it has sent into space, for just about every minute they've been up there. But only recently have scientists interested in people's problems on Earth begun taking an interest in NASA's research.

Ronald J. White was the chief scientist in NASA's life sciences division for twelve years, and he said the problem was that space capsules used during the first two decades of space travel were just too small and primitive to allow even the simplest of medical experiments. White said it has taken years for researchers to develop the new tools needed to do medical tests in space. And even Russia — which has a long history of operating space

NASA

Glenn said even he enjoyed the centrifuge training, which he did at Brooks Air Force Base in San Antonio, Texas. Glenn was strapped into the centrifuge and made two nine-minute runs that simulated the shuttle's launch.

stations, beginning with Salyut 1 in 1971 through the Mir station — hasn't probed medical research any better than the U.S. The orbiting space stations just weren't designed for it. Mir offered Russian, and ultimately American, astronauts "a marvelous opportunity to experience long-term spaceflight," White said. And some data has been gathered, both in orbit and by examining astronauts after their return to Earth. "But [Mir] was never designed as a research laboratory. It was designed as an outpost. . . . So it's not a kind environment for research work."

America had Skylab in 1973 and '74, and while it was a roomy orbital lab where medical research could be done, only three crews of three occupied it during those years, which limited the amount of data that could be collected. In 1979 it fell back to Earth.

It was just before that, early in the 1970s, near the tenth anniversary of Glenn's first flight, that it really began sinking in to the Columbus businessman who had been the redheaded space warrior that he never

would return to space. NASA programs were marching on, and he was getting old. During an anniversary interview, the Associated Press's veteran NASA reporter Howard Benedict remembers asking Glenn if he'd like to go back.

"I want to go back. I have this burning desire," Glenn said. "But Apollo is ending, Skylab is coming up, there's no chance. There's no way I can get back into it. I wish I could, but there's no way now. It's a dream, and that's all it is."

But Glenn was about to win that Senate seat he had twice failed to get, and it would not be many years before he took on a nearly invisible job: chairing the Senate's Select Committee on Aging. It didn't win him any leverage on Capitol Hill, but it put him inside the circle of geriatric scientists in America. ("I kid him that he's had a motive ever since he came in there," says friend Tom Miller, "but I couldn't say that that's true at all. But I love to needle him on that.") Coincidentally, this is when NASA began logging real strides in medical space research with the advent of the relatively spacious and reusable space shuttle, first launched in 1981. In addition to being a roomy, floating lab, the shuttles could be equipped with an extra detachable module snapped into place inside the shuttle's cargo bay. This module, called Spacelab, could be outfitted with specialized equipment for specific experiments that would not fit into the cramped quarters of the shuttle's middeck or flight deck. Although NASA planned the first Spacelab mission dedicated to life sciences in 1984, the Challenger accident in 1986 delayed everything. Still, White said the mission, for which he was the program scientist, finally flew in 1991. And suddenly NASA could see a line of advancement stretching between Mercury and Spacelab that was bringing a deeper comprehension of the effects of spaceflight on humans. "What we've been able to do is progress from an understanding of what the problems were that needed to be addressed, to very serious probes into what's happening and how," White said.

And then along comes John Glenn. He'd noticed what scientists already were curious about: the way lengthy periods of weightlessness look an awful lot like temporary versions of growing old. And now the man who at fifty felt the space program had passed him by after the Apollo era began seeing NASA in the space-shuttle age circle back toward him as a man in his seventies. His dream of returning to space might yet be rekindled with scientists wanting to understand why human bodies seem to react to weightlessness the way they do to old age.

"Who would ever have thought that circumstances would be such that age would be desirable for what we're trying to learn? Age would become an advantage in what we're trying to find out?" Glenn said, gleefully. "And that wasn't something I could have predicted, but I'm glad that it's occurred."

David R. Liskowsky, the NASA scientist in charge of the life sciences experiments on Glenn's flight, said Glenn developed an interest in the research at about the same time the science community did, and his full-court press to make himself a guinea pig for the research ignited the effort. "I think we were moving in that direction," Liskowsky said, but Glenn's overtures to NASA "to a certain extent did catalyze things. . . . Certainly the whole issue of studying spaceflight and learning about aging is something we're all very interested in, and something we'll continue to pursue after STS-95."

STS-95, Glenn's flight, would not be a Spacelab mission, but it would have a smaller, payload-bay module called Spacehab. Less elaborate than Spacelab, Spacehab would still provide extra elbow room and support equipment for experiments. Crew members would get to it through a tunnel.

Scientists came up with nine experiments to study Glenn's reactions to spaceflight. Only two were classified as "payloads," experiments that Glenn would actually do in orbit. The rest were "detailed supplementary objectives," measurements that would be taken before and after his flight.

Because of the difficulty of doing medical research in space, such before-and-after measurements taken over the years account for much of the knowledge scientists have about spaceflight effects on humans. "These DSOs have been done on a number of other flights, and we have a pretty large database, for some of them as much as forty, fifty, sixty astronauts. We have a pretty good idea of what a thirty- to fifty-five-year-old astronaut looks like on some of these measures," Liskowsky said. "Now we'll be getting that same data from Glenn."

The DSOs would address several effects that show similarities with aging, including balance problems, cardiovascular changes (especially blood pressure regulation), bone mineral loss and recovery, and effects on the immune system. The experiments Glenn did in space would look at two other areas where aging and spaceflight seem to have a lot in common: sleep disorders and muscle atrophy.

Astronauts don't sleep well in space, probably for a number of reasons. "The shuttle itself is kind of a noisy environment. The crew's obviously excited," Liskowsky said. "But we think there's a more physiological component."

It might be the same thing some scientists believe happens to us as we age: The mechanisms that regulate our internal circadian rhythm — the "body clock" that makes us get sleepy when night falls — get out of kilter. Research indicates the natural cycle of light and darkness sets our clocks, but something seems to disrupt that pattern as we age. Astronauts in space may suffer the same problem, but for a different reason. In their case, the culprit might be the compellingly beautiful view out the orbiter's windows. "When astronauts finish their work day, the best show in town is out the window," said White of the National Space Biomedical Research Institute.

But that exposes them to extremely rapid changes between daylight and darkness. "You're circling the Earth every ninety minutes," Liskowsky said. With the sun popping up from behind the horizon once every orbit, it's like telling the body it's morning sixteen times a day.

NASA has used bright light — a treatment actually designed for aged people — to "reset" astronauts' body clocks before a mission.[1]

Originally, Glenn was supposed to be a central test subject in another possible sleep treatment, the popular and controversial substance melatonin. Melatonin is a hormone produced in the pineal gland, a pea-size organ in the brain. Melatonin production coincides with darkness and sleep, and studies have shown that small doses can induce sleep and affect the sleep cycle.

The experiement had called for having Glenn and Dr. Chiaki Mukai take tablets at the start of each sleep period during the flight. Some would be melatonin, and some would be placebos — they wouldn't know which was which. But about two months before liftoff, NASA doctors discovered that Glenn couldn't take melatonin because he failed to meet one of the medical parameters approved for the study. Doctors as well as a spokesman for Glenn insisted the hormone posed no danger, but they cited medical privacy as the reason for not explaining what disqualified Glenn from the study. Liskowsky said losing melatonin data on Glenn was a minor loss. The matter would erupt into a brief controversy when the decision, concealed for months, became public just weeks before the flight.

But Glenn would remain a primary test subject in all other aspects of the sleep study. Mukai would still take the melatonin, and Glenn would not, but together they would remain the focal point of evaluating how restful space sleep is. They would swallow husky tablets containing tiny thermometers that measure the core body temperature, one measure of the circadian rhythm. Also in each pill would be a radio transmitter that would send the temperature data to a receiver Glenn and Mukai would wear on their belts.

On four nights — two early in the mission, two late — they would don elaborate electrode nets that would measure everything from eye and muscle movement, to breathing rates and brain activity. The following day, they would take tests aimed at measuring cognitive performance.

The other study central to Glenn's flight would examine protein turnover during spaceflight. Another effect of spaceflight is muscle atrophy, a loss of muscle tissue result-

NASA

Glenn borrowed the commander's seat in the cockpit of the space shuttle Columbia, but he was assigned the tasks of a payload specialist for the Discovery mission.

ing from decreased protein production.

Something similar happens as we age. The big difference is that astronauts quickly recover after a spaceflight; the change in aging is permanent.

This experiment would involve Glenn and the shuttle's rookie spaceman, Pedro Duque. On two different mornings, they would take a dose of two special amino acids, the building blocks of protein. The first, alanine, would be swallowed in pill form. The second, histidine, would have to be injected. For the next three days, they would periodically take blood and urine samples for analysis. The amino acids would carry a nonradioactive marker that scientists would use to trace their distribution in the samples.

The loss of muscle was long thought to result from the zero-gravity environment: Astronauts don't get much exercise from

BILL REINKE

Glenn's return to space quickly became a major news event.

floating around in a spaceship. But Liskowsky said research has raised some doubts. "People thought if you just exercised that would keep the muscle up . . . you wouldn't lose the muscle," he said. But, "even with exercise, though you can improve, or decrease, the muscle atrophy, you still see some muscle atrophy in astronauts. . . . We don't know whether they're just not doing the right exercise, or maybe there's something else."

Liskowsky said the protein experiment was set up to find out whether "something else" might be a flood of stress hormones that cause or influence the breakdown of protein. Similar effects have been seen in burn patients who lose muscle despite extra protein in their diets. Could astronauts be losing muscle because of stress? Despite their calm, cool voices, odds are they experience just a tad of stress from riding a skyscraper-size rocket into space, being responsible for batteries of expensive and complex experiments, then blazing down through the sky like a meteor. It isn't clear what ties that might have to aging, but Dr. Arny Ferrando, the University of Texas scientist in charge of the experiment, said he was happy to have an older person to study. "From this one study, I won't make sweeping generalizations . . . but I'm interested," he said.

Bone loss is another result of spaceflight,

but Liskowsky said there were no on-orbit experiments to study the process in Glenn or his crewmates. That's because a nine-day mission wouldn't be long enough for much bone loss to happen. "Given any mission, you're limited to how much time you have. We wanted to do experiments where we expected to get a larger response," Liskowsky said. "The muscle atrophy happens quicker. . . . We do expect to see a fair amount of muscle atrophy." Still, measurements before and after the flight would record the amount of change in Glenn's bone mass.

So what would scientists learn about aging from Glenn's flight? Nobody knows, Liskowsky said, and that was the point. "We know what he looks like before the flight and how he compares to the younger astronauts before the flight, and now we'll see what the changes are after the flight. Has he changed the same amount, or has he changed double the amount, or maybe less? We don't know, and those are the questions we're asking."

Glenn's experiments weren't expected to solve any mysteries. Even NASA chief Dan Goldin played down the results the agency expected from them. "I don't think we're going to wake up on January first, evaluate the data and say, 'Voila, we know how to solve sleep disorders.' I think what we may have is some recommendations on how to take the next step."

The two on-orbit experiments would not be Glenn's only duties. Liskowsky said Glenn would have plenty of work to keep him busy: "The life sciences experiments that we've been talking about only make up about ten percent of the whole mission. . . . He has responsibilities like all the crew members for other things going on in addition to his participation in medical experiments."

Every time he met with the press, Glenn tried to shift the focus from himself to the mission's science. After all, the hero of the space race would not be blazing a trail to the heavens this time. That trail has been blazed, cut, and worn into a path since his first flight. Discovery would be making its twenty-fifth trip into orbit, and it would be the ninety-fourth spaceflight in the U.S. shuttle program. "I fit into a whole new role now," he said. "I'll be doing more experimenting, something that's more technically scientific perhaps than the observations and impressions I had on that first flight."

But it made no difference. In fact, Glenn seemed to be dogged by the same problem that has afflicted the space shuttle for years: The public didn't give a hoot about all that science stuff. Glenn's body might yield the first in a long string of data points that someday increases our understanding of the aging process, but the prospect excited few people outside the medical profession. The public wanted a hero, and John Glenn would be one whether he liked it or not.

NOTES

1. Joan Vernikos, Ph.D., Director, NASA Life Sciences Division, "The Study of Human Adaptation to Space Helps Us Understand Aging," NASA shuttle Web page.

SCOTT PARAZYNSKI

PEDRO DUQUE

PEDRO

COMMANDER CURT BROWN

NASA

GLENN

CHIAKI MUKAI

MUKAI

STEVE LINDSEY

STEPHEN ROBINSON

CHAPTER 10
THE FINAL STAGES

East Texas still swelters in mid-October and a balmy breeze tugged at the rows of star-spangled John Glenn banners along NASA Road 1 on the southern skirt of Houston. The road, and the banners, led from the flat highway that drives north into the city's shimmering downtown straight to the low-slung campus that is the Lyndon B. Johnson Space Center. Mission Control.

The city of Webster, a small town that shares its city limits with Houston, used the banners to rename its stretch of NASA Road 1, The John Glenn Parkway. At least for a few weeks. Discovery's launch was now two weeks away and the celebration had begun. In recent weeks Glenn's face smiled off the slick covers of *Time*, *Newsweek*, and *Life*. Television networks rolled out waves of interviews, video clips, and astronaut experts who, on launch day, could tell it like it really is. It had been a while since a shuttle launch made the cover of *TV Guide*. The merchandising arm of NASA was pumping out every John Glenn trinket it could dream up, from T-shirts declaring "Once Is Not Enough, Do It Again John Glenn!" to commemorative plates to pewter baby spoons with the Discovery mission logo etched into the handle. Even The Weather Channel planned a special on how bad weather delayed Glenn's Mercury launch in 1962.

All these things amounted to symptoms of a nation's fever over one man's triumph, but a triumph over what? Age? Yes, age. And odds and naysayers, too. But perhaps most remarkable was his victory over the gravity of his own life story, which surely insisted that his most glorious day was long behind him, back then at the beginning of his life when he jogged instead of "power walked" and had a head of red hair that was not yet overpowered by white. Yet blended into the material celebration in the thrill of the new John Glenn space ride lay a quandary for us all. For NASA, the press, even America and the world: Do you revel in John Glenn's return to space for its historic novelty, essentially playing the hero card and tipping the game away from six other crew members and their eighty-three science experiments? Or do you restrain yourself, take this opportunity to embrace the space program and all other astronauts, and leave out the part about a seventy-seven-year-old man who used the sheer force of will to recreate a magical period in his life and that of his country?

Tough call.

The John Glenn banners snapped in the air along NASA Road 1 about 100 yards from the Johnson Space Center auditorium where Glenn and his crewmates, in their last press conference before launch, talked

TY GREENLEES

Senator John Glenn put his arms around fellow astronaut Dr. Chiaki Mukai after a reporter asked about the personal relationship between the two STS-95 crew members during a press conference with the astronauts near the launch-pad emergency escape system.

earnestly about their scientific mission. The crew of seven — two Air Force fighter jocks, one intense scientist, a young doctor who dreams of Mars, a computer whiz from Spain, a rambunctious cardiologist from Japan, and the aged space hero — drew a packed house and lit the room with their enthusiasm for the work they do. They love it so much, not one of them didn't go on a bit long in summarizing his or her work aboard Discovery. Dr. Chiaki Mukai, the cardiologist, drew hoots of laughter using a plastic plant with construction paper roots to explain her cell growth experiment. And Stephen Robinson, a Stanford Ph.D., showed why the rest of the crew called him "Steve Ney the Science Guy,

"with his lengthy run through a multitude of topics. Even the pilot for the mission, Steve Lindsey, an Air Force lieutenant colonel, grinned broadly in noting, "We're setting a new record on this flight with twenty laptop computers" to run all the experiments.

And then Glenn, at the end of the table, last to go, beamed a pride that could have been grandfatherly if he hadn't been working so hard to be a peer. He gestured back down the table with that right thumb he has used for a million photographed thumbs-ups, and said: "These are brilliant people here! I'm not just blowing smoke at the folks to my right here. They're brilliant people who are putting these things together and doing the job that's representative of everybody, and I just wish there was the excitement about every flight that we used to have when we were doing some of the first [launches] way back in the early days."

It was the second time in a week that Glenn had blasted off on the science vs. hero issue. A week before, the crew was in Florida, at the Kennedy Space Center launch site, running through a countdown test that involved a complete practice of launch day right down to strapping into the shuttle. The crew stood in a sandy pit with their backs to the towering launchpad gantry that wrapped the Discovery in metal, like a lone beach-front high-rise still in the steel construction phase. Glenn made headlines with his brief pique.

"I'm going to castigate the press here a minute," he began. "Too often you get into the human aspects of this, you don't get into the scientific stuff that gets into everybody's household over this country. This program and what we're about on this with these eighty-three experiments, most of them are designed to benefit people right back home. When you see people sitting there watching the TV at night with their kids there, they're benefiting from this kind of a program. And we have big acronyms on this. . . ."

Glenn fished from his jumpsuit pocket a piece of paper, onto which he had scratched the details of the Biological Research In Canisters or BRIC program that Dr. Mukai would do on plant cells. He also read about the Microencapsulation Electrostatic Processing or METS experiment, in which scientists are trying to perfect a medicine capsule that could be carried in the bloodstream to release drug doses directly into cancer tumors. "So I wanted to take just a couple of minutes to point out; I hope you can get into the science of this flight, I hope you get into the eighty-three things. . . . I know it's easy to report the 'oh-gee-whiz' of the personal aspects of this thing,

TY GREENLEES

Senator/astronaut John Glenn shares a moment with his wife after his arrival with the Discovery crew at the Kennedy Space Center for the launch that would return Glenn to space.

but this is science at its very best out there on the cutting edge. And I'd sure invite you to get in there and report that."

A second later, before the next reporter's question, he added: "I just got fired up on them. I haven't even got started yet."

If only it were that simple. If only getting fired up and giving lectures about the truly fascinating science, with its real-world, Earth-bound applications, could dim the appeal of John Glenn's return to space. If only John Glenn were a hero in some other realm. Maybe a cowboy or movie star. Heck, if he'd succeeded as president instead of succeeding among the Mercury Seven, maybe things could be different. But what complicated the reaction to Glenn at the table back in Houston, wagging his thumb to direct attention at his crewmates, was the indelible truth that he was their hero. Anyone who has made

Glenn answers questions from the media at the Kennedy Space Center where STS-95 Discovery crew trained. The space shuttle Discovery is on the launchpad but is covered by an enclosure.

a career of spaceflight, either in the flying or in the engineering of it, is steeped in the Mercury story, and therefore in the John Glenn story. He is why I'm here, astronauts around NASA would say. And this particular crew, the six other members of Discovery's flight crew, was no exception. They were babies or children when Glenn first flew, and they never forgot the power of what he did, even if they really only learned about it by reading the history years later.

Dr. Scott Parazynski, a physician on Discovery, has never wanted to do anything as much as fly in space, and he typified the circumstance. He was seven months old for Glenn's first flight, and while he gushed at the prospect of learning more about the sun by deploying the Spartan 201 satellite during the coming mission, he opened his remarks in Houston with this: "Most of all, I'm thrilled and excited to share the sky with one of my boyhood heroes, John Glenn."

Even Commander Curt Brown, the man who shared with Glenn the burden of press interviews and the effort to redirect them, and who as the final word on media access during training fought endlessly to keep the crew on schedule, also couldn't resist the draw of history. "Nowadays, we work in space, so we take it for granted that we can get in the shuttle and get to orbit and do the science. . . . [The Mercury astronauts] had to prove that we could get to orbit. I think they deserve the hero status a lot more than we do, and I'm just lucky to have a job. I think most of the crew would agree with me, that we're just lucky to have a job where we go to work in space."

The joke among the crew was that you couldn't swing your arms in training without knocking into a cameraman. Not only did the outside world want in, NASA also was recording every breath, having brought in a photographer from *National Geographic* who all but worked on NASA's staff and had

to give the agency the bulk of what he shot. In exchange, though, he and his magazine got to keep their favorites for publication.

"The one rule I had is it could not impact training, it had to be invisible to the crew training wise," Brown said. "There was always the bottom rule that says if it became an impact during the event that I could pull the plug on it and we'd cancel and ask the folks to leave. . . . See, we want to talk as much as we can about the flight, but we also want to train right. And we can't let that interfere with our training because we don't want to go up there and make a mistake or not be efficient because it takes so much effort to get there. So, we walked that thin line, and I think we've done a good job. And we're ready to go fly. . . ."

The senator left his day job for good in the middle of the last week of September. You never would have guessed. Congress was just coming out of its paralysis over shockingly graphic details of President Clinton's affair with a White House intern, and momentum was growing to get the must-do spending bills passed quickly so people could go home and campaign. Glenn, the senior senator from Ohio, was increasingly booked with NASA duties. Since Congress' regular summer recess in August, Glenn had spent more than half of his time away from Washington on training exercises, and by this week in September he was about to make that full-time. Unless a controversial bill of epic proportions demanded his return to the Senate, for the month of October, the month of the launch, he belonged to NASA. So on a Tuesday evening, September 29, Glenn cast his final vote on the Senate floor. (It was in favor of a procedural motion on a bill to put a morato-

TY GREENLEES

As always, Glenn tried to take the focus off himself and lauded his colleagues aboard Discovery as an impressive bunch.

rium of state's taxing on Internet sales.) Glenn noticed this was, for all intents and purposes, the end of his Senate career, but he didn't dwell. Too much to do.

"Yeah, I thought about it," he said the next day. "I'm more proud of my voting record in general, all those 9,414 [recorded] roll call votes, than I am of anything else in here. Because I think when I came in here, I was concerned that some of the fringe political elements, I felt, were on the ascendancy then, and I wanted to represent more than anything else the two-thirds or three-quarters of people in Ohio who don't see themselves as being on the political fringes. They just see themselves as hardworking people who go to work every day and their views are pretty much out there in the middle of things, sort of in the center of the

political spectrum. And I think my voting record would reflect the views of those people."

Not a bad salutation, but he didn't say any of this on the Senate floor for the record, or for history. He said this sitting in his high-ceilinged Senate office with a small group of Ohio reporters. Officially, for the record, for history, John Glenn left the Senate without a peep. A rare confluence of enormous responsibilities — preparing for a spaceship ride at seventy-seven and putting the customary coda on a lengthy career in the United States Senate — had led to something even rarer still: John Glenn left something undone. Not a big thing, in the scheme of it all, but something he had wanted to do.

"I had planned to make sort of a farewell

speech on the Senate floor, [but] I've been so busy with everything else I just didn't have time to sit down and put it all together," he said. "And I did want to do that, and that's one thing I regret. . . ."

In fact, he had become so swamped with shuttle training that when his Democratic colleagues in the Senate staged a retirement dinner for him and two others who also came to Washington in 1974, Glenn couldn't go. He sent his wife, Annie, and then appeared live via satellite link from a NASA studio in Houston, like some out-of-town movie star thanking the academy for his Oscar. Of course, Glenn wasn't off getting a tan in the French Riviera; he was knuckling down, doing one of the many unpleasant things a scientific test subject must do. "I couldn't be there because [that] week down there at Houston was the base-line data week, and that means I was giving blood two, sometimes three times a day. And urine collection and . . . everything, doing analysis of everything."

The other thing Glenn would not do, it was now clear, was help the Democratic Party try to retain the seat he was retiring. Ohio's popular Republican governor, George Voinovich, had declared for the seat early and had a mountain of cash to run against a far-outmatched former county commissioner from Cleveland who had neither the money nor the name recognition to seriously challenge him. But Glenn, with perhaps not as much regret as missing the chance to say goodbye on the Senate floor, declined even to endorse his party's candidate, steering clear of politics the way other NASA flight crews do. "They stay away from that, and I wanted to fit in with them down there. I thought it was important that I do it their way, and that's it."

He felt bad about that, too, he said. But the absolute world of oxygen bottles and emergency egress training, of days scheduled in five-minute increments and seven million

pounds of rocket thrust, what NASA's Administrator and then Discovery's commander called the "unforgiving" world of space travel, had taken priority. Had to take priority, no two ways about that. So there, in an instant, Glenn shed the dark suit and shrewd equivocations of the political man and strapped into the concrete intensity of the modern astronaut. Calling this transition seamless almost doesn't cover it. "Yeah, I'll miss it, sure," he said of the Senate. "But, you know, move on to other things, and that's it."

Or maybe move back to other things. No matter what he said about feeling at home in the Senate, which surely he did after two decades, it instantly was clear that home could be one kind of place while heart was entirely another. If you dropped in on Glenn just at that moment, in time to watch the switch, you'd swear you were looking at a space program lifer coming back to the job after a long, forced departure. You'd swear you were seeing a man come back to himself.

John Glenn the politician loved his white-collar duty, but it didn't make his heart beat like the blue-collar business of risking your neck.

Glenn's daughter, Carolyn Glenn, who goes by Lyn, always noticed the way her dad connected with people who bargained with their lives to fulfill their duty. Whether it was stopping to chat with the Marines standing guard at a U.S. embassy he visited while abroad, or sharing a mutual respect with the Secret Service agents assigned to his campaign during the 1984 presidential race, he spoke their fearless language.

"I've seen him in the Senate be just extremely excited about certain votes," she said, "and he'd really work on getting the appropriations to go through and he'd really have worked on it for weeks at a time, and all the ups and downs and the different personalities, but it's not, that's not the kind of,

TY GREENLEES

Glenn prepares to drive the M-113 armored personnel carrier with instructor George Hoggard (left) at the Kennedy Space Center. The astronauts drove the M-113 as training for a launchpad evacuation.

that's completely different from this.

"This is incomparable."

"This" being his shuttle flight, and this being him.

"He felt an affinity with the Secret Service, and I saw it, there was an affinity. Agents asked to be on his detail," she said. "Dad knows what it's like to literally put your life on the line for your country. And those agents know the same thing because that's their job. They are literally willing to die every day that they go to work.

"So it's that kind of fiber, or core of a person that is only brought to full fruition very infrequently. It's not the common path that most people take."

Which raised again the question, is it science or is it heroics? Even within Glenn was this elemental debate over where to find the true engine driving his return to space.

Today's astronaut is one part curious research scientist and one part thirsting space pioneer, which is perhaps not an inaccurate description of Glenn, too, even if he is a product of the earliest era in space exploration when a healthy dose of swagger was mixed in with every astronaut. But which comes first in him? Annie Glenn had been saying lately that her husband would have been a doctor if the war hadn't come along during college, but you'd be hard-pressed to imagine him being any more adept at anything, including medical research, as he was at being a fighter jock. Could it be both? Could it be that Glenn was simultaneously driven to conduct test-tube research, and lift off from a concrete pad for the sheer joy of mastering the unforgiving world of space?

Maybe. Lyn, fifty-one, had to think for a minute to recall a time when she remembered him being this happy. Not that he's been unhappy, but the last time he felt like this. "On his seventieth birthday," she finally said, "when he water-skied behind his boat, he just thought that was the cat's meow. That's different from this, but it's that same feeling of great satisfaction, of 'Boy, I can do this!'"

So why couldn't he just say that now? Just talk a little about the personal satisfaction of it all? Well, it's worth noting that he couldn't say it in 1962, either. Glenn splashed down to a worldwide celebration, but he was far less interested in taking bows than he was in making the point that his flight proved the need to have people riding in space capsules. At a time when many argued that flying humans was needlessly risky because instruments could measure everything we'd want to know about the space above Earth's atmosphere, Glenn said the "most satisfying" part of the flight was him having to take over control of the capsule when the automated system failed. No instrument could have done that, and if not for having a pilot aboard, the capsule would have been lost. And again, too, in 1957, Glenn never missed a chance to say that his transcontinental speed record was not just a stunt, or even man's audacious effort to advance his mastery of aviation, but rather he insisted that it was an important, incremental performance test in the evaluation of the jet he flew, the F8U Crusader. It's as if he signs the autographs, but doesn't like to think about why people ask for them.

"If he could embrace that more he wouldn't be the person is," said Lyn. "Because his ego would be very different. He is very much the person you see, and that's very much a dichotomy because you see and you want more. But if he gave you more he wouldn't be the person you want to see."

The result, with his space shuttle launch closing in, was more of the same appeal to science Glenn had been making since the announcement in January. But in these closing weeks, Glenn's wife and daughter at last began talking, and what they said brought into steeper relief his march-ahead fervor. All along, Glenn had acknowledged that his family was "cool" to the idea of him going back up, but all along he swore he had won them over with the import of the science. "We had talks about [safety], they questioned the whole thing," he said. "But their view on it really has changed. The more they've learned about it, why the more they agree that it's something we ought to do."

Not exactly. When she finally was speaking for herself, Lyn Glenn talked about "acceptance" of her father's latest mission, of having to work hard to reach a point where she could stop opposing it outright and come to some terms that let her live with it. She didn't like the idea when he first raised it with her over the holidays in 1995. "I said, 'I'm against it. I understand that you're curious about it, and I sure get that, but I really don't want to go through that again.'" And she never reached the point where she actually favored it. "I guess what I know is that I will support the people I love to be who they need to be. And I may not agree with them, and I may wish they would do things differently, but I am powerless over their decisions," she said. "And yet I can stand back and value that my father is being who he needs to be."

Son David Glenn, fifty-three, was not talking, but Lyn said, "Oh, Dave's very clear about it, as I am. But he has done his own work around acceptance of this."

What really brought them around, it turned out, was not so much the science — though they all were fascinated by its potential — but the effort of dear old dad. They may or may not have been convinced that the experiments were "something we ought to do" no matter the risk, but from the beginning they clung to their skepticism over

Like all pilot astronauts, Glenn's flying skills were tested in the T-38 training jets NASA uses to transport shuttle crews.

whether he could perform the job of an astronaut at the age of seventy-seven. Until he himself put it to rest. As Lyn put it, "I was certain that his enthusiasm could take him a long way," but she wanted to leave room for reality to make its case. She waited until NASA decided if it was interested — it was — then she waited to see if her dad could pass the invasive and rigorous physical exam required — he did. From there it was a matter of seeing how well he stood up to the day-in, day-out of preparing for a space shuttle ride. Of course, he handled it with the steely persis-

tence of any combat pilot or Buck Rogers rocket rider, mixing a focused mind with a courageous heart. He handled it like any hero, if you dare.

"I knew John's mind was very, very good," said his wife, Annie. "And his memory and everything like that. And I also knew John was very strong and able, but after being [in Houston] two weeks, I went home completely excited because he was so excellent. I mean, the difficult aspects of what he was doing were just tremendous, and I think a lot of the younger ones were surprised at him, too. And also, his mind was just taking it all in. I was really excited and proud."

Lyn had gone to Houston from her home in Minneapolis to spend the final weeks with her mother and father before he returned to space. She stayed with them and watched her dad grind out a schedule that began at 6 a.m. and ended when he "kind of falls into bed about 10 or 10:30 or so. . . ." In addition to fulfilling his training duties during the day, Glenn spent his evenings signing autographs or responding to letters and phone calls.

"I think he is at absolute peak life right now," Lyn said. "At peak time. If I was doing what he's doing, I would say I would be exhausted. But he is not exhausted, he's exhilarated."

Glenn squinted at the sunlight of a perfect Florida day. October 26, 1998. The day the actual countdown to his shuttle launch began with the Discovery crew's arrival at the John F. Kennedy Space Center. As is NASA custom, they arrived from Houston in T-38 training jets, two-seaters, and Glenn rode in the second seat of Curt Brown's plane. As was Glenn's custom, he drifted out to the microphone planted before a thicket of news cameras and reporters making sure he stayed several steps behind the rest of the crew. His fingers laced his hands together in front of him, the kind of thing you do when you're not sure exactly what to do, but you're positive you don't speak first. Payload Specialist 2 is down the totem a ways.

"We're here," Commander Brown said to the crowd, "We're happy."

That could not have been any more true than it was for Glenn. Training was over. Every flight simulation had been run, every experiment briefing paper absorbed, studied, locked in tight. The eight-month process of taking in

all there is to know for any given shuttle flight, what astronauts say is like drinking from a fire hose, had mercifully ended. For the last four days, the crew had been in quarantine, which meant that anyone who got close to them must first have passed a physical exam. They'd stay there until launch.

"It's time to wind down a little bit," Stephen Robinson, the science brain from Stanford who was on his second mission, said. "Going into quarantine, where you kind of withdraw from society . . . it's sort of, you get halfway into space. It's a great thing to do; it's just the best thing. We're all trying to get as much work done as we can so when we get to quarantine we can just review our notes. Don't gather any more, just review information you already have."

The only worry now was the abrupt development of a ferocious hurricane inching about off the coast of Honduras. Hurricane Mitch was a stout Category 5 storm with winds of 180 m.p.h. and such a well-defined eye wall that it looked like the bleachers in a stadium, and raised the possibility that its powerful, low-pressure system could generate winds fierce enough, from 8,000 miles away, to delay the launch. But the potential for bad weather, which had played such havoc with Glenn's first launch those years ago, couldn't temper the buoyant mood of the crew, or Glenn.

"I'm John Glenn, I'm PS2 on this flight, and very glad to be here," he said when it was his turn. "And one word on this whole thing. I have been pleasantly surprised at the outpouring of interest in this flight. And it's really gratifying to see people get so fired up about the space program again and about their interest in it. This is going to be a very research-rich flight, we've got about eighty-three different projects. . . ."

Senator John Glenn talks with the media after his historic flight aboard the space shuttle Discovery.

CHAPTER 11
THE RETURN

When the countdown clock on the lawn at the Kennedy Space Center reached zero and the rockets ignited, the clock immediately began counting back up again. The numbers climbed as Discovery did, recording what NASA calls Mission Elapsed Time or M.E.T. And although the numbers were advancing up from zero now, rather than retreating down toward it, the passing of M.E.T. represented no less of a countdown than the five-four-three-two-one of liftoff. The timer now was moving toward something as immovable and reliable as zero; the inevitable question that greets the conclusion of every great achievement. What next? For John Glenn and his glorious ride back to the forefront of American pioneers, the question would come at him from two levels. There was, of course, the grand question about how he planned to live his life now that he had resolved the quest that ached in his heart for almost four decades. But also for Glenn at the end of Discovery's trail would be the painfully finite issue of whether the seventy-seven-year-old man could get up and walk off the shuttle when it stopped rolling on the runway. It would be hard to say one of these was more significant than the other. Space travel can level a person. After days and days of needing to expend only the slightest effort to move, of having utterly no resistance to even the simple

act of standing still, the body gets used to the "holiday," as Discovery's Dr. Scott Parazynski called it. And the return to a world where an invisible force is forever trying to pull you to the ground can be quite strenuous. Getting into space may actually have been the easy part after all, compared with the peril Glenn's steely image faced in the prospect of needing a stretcher at the end of his journey.

But through the long flight, those questions were suspended. There was so much to do. Not just the eighty-three scientific experiments — although Glenn was doing his best to make sure they weren't forgotten in the public eye — but Discovery's schedule in orbit also was wrapped up in the John Glenn story. Press conferences abounded, questions and answers with school kids made the news, and Discovery even did a few minutes with the "Tonight Show's" Jay Leno. When Glenn chatted with Johnny Carson back in the day, it was Glenn who stole the show. Not this time. It was Commander Curt Brown. The Air Force lieutenant colonel who had managed such a perfect balance between taskmaster and cruise director in preparing his shuttle team to be both efficient and relaxed, also carried in him a heavy dose of the quipster.

"This is the most amazing thing that has ever happened to me," Leno said in opening the radio conversation. "This will help me get

BILL REINKE

The shuttle turns into a heavy glider when it comes back to earth.

the final chapter of my sixth grade book report on Senator Glenn. I had to write a paper on the Friendship 7 in the sixth grade, and I can turn it in now; it's complete."

Then followed the brief Q&A where Brown shined:

Leno: Does Senator Glenn keep telling you how tough it was in the old days, how cramped it was, how small it was, how lucky you young punks are?

Brown: Well Jay, actually, no he doesn't always do that. Only when he's awake.

Leno: Who's driving [the shuttle] right now?

Brown: Actually, Jay, it's just like California; no one's driving.

Leno: Now, what can you see on Earth from space there? Is it true you can see the Great Wall of China?

Brown: Well, Jay, sometimes, if the lighting is good we can see the Great Wall of China, but we just flew over the Hawaiian Islands, and we saw that. And Baja, California. You can see the Pyramids from space, and sometimes rivers and big airports. And actually, Jay, every time we fly by

California we see your chin.

Even Leno sounded like he was caught flat-footed by that one. Everybody enjoyed the laughs. After the radio link with Leno ended, mission control praised Brown, Pilot Steve Lindsey, and Glenn, who together had matched wits with one of America's premiere comic minds.

"Thanks to you, the crew of Discovery. You guys are real comedians," mission control said. "If you ever get tired of the space business you've got a bright future."

But Brown wasn't done yet.

"Now that we're fired," he said, "can we stay up an extra day?"

"Stand-by," the radio voice sighed, "we're checking. . . ."

Jokes aside, work bracketed the sliver of time spent with Hollywood. Glenn ate right after the Leno interview, and then Discovery prepared to take some photographs of Earth. The shuttle was just passing diagonally across Mexico and soon would be crossing Central America, where Hurricane Mitch — the storm that was just an annoyance in threatening to delay Discovery's launch days earlier —

BILL REINKE

The shuttle touched down smoothly at Kennedy Space Center.

had charged ashore in Honduras and killed thousands. "Clouds should obscure Yucatan," mission control messaged the crew, "but hopefully will clear out over Nicaragua and Honduras. Great interest in photography over this devastated area." And earlier that same day, the Discovery crew had successfully completed one of the biggest goals of the mission: retrieving the Spartan 201 satellite. A tiny spaceship that weighed about as much as Glenn's Mercury capsule, the Spartan was released from the Discovery payload bay to spend two days taking readings of the sun. The delicate reunion of Discovery with Spartan, almost 340 miles above the Earth, ended with Stephen Robinson gently snaring the satellite with Discovery's robot arm. Glenn was in the Spacehab, the extra module snapped into place inside the payload bay to give astronauts extra room to work, looking out the window. He had a camera, and the best seat in the house.

"I was within about fifteen feet of it when the arm grappled it and started bringing it back in," Glenn gushed to Walter Cronkite during a radio interview. "It was

just a fantastic sight."

They all were fantastic sights for Glenn. In 1962, he was one of three men on Earth who had seen the planet for what it truly was, a magnificent blue marble suspended in dark space. And now, while hundreds had seen what he first saw, the picture out the window still was the inspiring spectacle he remembered it to be.

"I don't think you can be up here and look out the window as I did that first day and look out at the Earth from this vantage point . . . and not believe in God," Glenn said during a radio press conference. "[It] is to me impossible. It just strengthens my faith, and I wish there were words to describe what it's like to look out the window from up here and see some, I guess we're looking at about a 4,000 mile swath of Earth go by under us."

And in the days while Glenn was aloft, Annie did something remarkable, too. She held a press conference with her two children in the auditorium at the space center in Houston, talking easily and at length about how the family was dealing with the shuttle flight. This marked a significant difference

153

BILL REINKE

Glenn spent about five hours in orbit his first flight, and spent nearly nine days aboard Discovery by the time the shuttle slowed to a halt.

between past and present, as big as any development in spacecrafts or pressure suits. In 1962, when the world clamored for every tiny detail about America's courageous new space man, Annie, the space man's wife, was silent. Captivated by a severe stutter, Annie always had others speak for her — answering the telephone in the house, ordering dinner at restaurants — so the idea of pulling up a chair on a stage before cameras and reporters was beyond the furthest reaches. NASA's Manned Spacecraft Center, as it was called at the time, put together a booklet about Glenn and his first flight back then, and it quoted Annie when she appeared at a news conference with the other astronaut wives. She didn't talk, but rather blipped out minimum answers to questions directed at her.

Would you like to see John fly again?
"Sure."

Do you think son David would like to be an astronaut like his dad?"
"He knows a lot about it."

With the whole world watching, it was a difficult time to have no tongue.

"People decided that I was shy," Annie said during her shuttle press conference. "I was not shy, I just couldn't talk."

But a speech therapy program in the early 1970s finally tamed her stutter, making possible a day of honesty and pride during the mission of Discovery that simply wouldn't have otherwise existed.

"The difference now, to be able to talk to all of you," she said. "I'm really touched by all of you being aware and wanting to come to hear Lyn and Dave and me. And that I can talk to you, that just means an awful lot to me."

What she said, of course, was powerful, too. That she cried on launch day (but not until those volatile solid rocket boosters had been used and safely jettisoned), that she couldn't wait to touch her husband again, that it felt good to see her lifelong companion so deeply and genuinely happy. She said they had revived their goodbye ritual from his combat days, the one where he says he's just going to the corner store for gum and she admonishes him to hurry back. It had been a helpful bridge over uncertain waters for John and Annie, something that hadn't been so important since they did it by telephone while he was strapping into his Mercury capsule in 1962.

Some wondered if Glenn could walk off the shuttle after so much time in space. He did.

BILL REINKE

But this time, John Glenn had done it one better. On the night before the launch, he gave her a pack of Doublemint chewing gum, which they agreed she would save for his return. "We're going to chew it together," she said.

Also talking at this press conference was David Glenn, who had chosen not to talk before. His silence through this shuttle flight was more self-imposed than Annie's had been in the Mercury days, but it wasn't any less absorbing when he broke it. He needed little prompting to expand on the clip NASA television had broadcast of him expressing his grave unhappiness with his father's return to space.

"When we first heard about this idea, it was pretty hard to accept. I had an image of him and my mom getting out of the Senate, having time to do the things they'd talked about doing. . . . What immediately came to my mind was the whole Challenger thing, because I'd watched all the films of that 100 times at least. I think, like a lot of people, I had that reaction, I just sat there staring and watching it over and over and over, and it was such a shock.

"And I knew that everything had been going very well and statistically it had been an incredibly safe program, compared to what people might have guessed at the beginning, early on. But it took a while to really adjust to this. My initial reaction was really being just mad. But . . . I got a couple of books that went into a lot of detail about what the shuttle program is, the details of how the shuttle works. That sort of made it really feel like this is a tried and true machine that works well, and I knew the people and I just really relaxed.

"But when it came down to watching that launch, that was really scary. That was very scary. Once the boosters dropped off . . . seventy-five percent of my anxiety was gone at that point. And when they were in orbit . . . I haven't been worried since. I'll feel a whole lot better when he's back on Earth, but I think at this point I haven't been losing sleep. It's great to see him up there having a good time."

From liftoff, the M.E.T. of the Discovery flight — John Glenn's return to the celestial frontier — ticked incessantly until the thud of tires on a sun-warmed runway halted the numbers at 8:21:44:54.

Eight days, twenty-one hours, forty-four minutes and fifty-four seconds.

In the control room, the landing came with no less tension than the liftoff, which had been only minutes away from being scrubbed for the day because a few curious pilots flew too close to the pad at a critical moment. For the landing, NASA's engineers in Houston were eye-balling a weather situation that was threatening to disrupt things. Like the buzzing planes of launch day, this was not a problem of severity, but a problem of timing. Hurricane Mitch, which had briefly worried engineers before the launch, returned as a tropical storm in the days leading up to the landing, creating an atmosphere of uncertainty. Fortunately, it blew across Florida and harmlessly into the Atlantic Ocean a full day ahead of the scheduled touchdown. But then a cold front pushed down into the Sunshine State, and suddenly on the morning of the landing a strong breeze was sweeping across the Florida runway, edging up to sixteen or seventeen m.p.h. With gusts of seventeen, NASA wouldn't land at Kennedy. When the shuttle comes in to land, it is under no power of its own.

Rather, it's a big, heavy glider, which makes it more vulnerable to wind conditions. A spiking cross wind can make the shuttle tough to control, so NASA set the seventeen m.p.h. limit. Edwards Air Force

Elementary school students and NASA families ask Glenn for his autograph during a welcome home celebration for the Discovery astronauts in Houston.

Base in California was the original landing site of the shuttle program and made a fine backup for Discovery, but the preference was Kennedy, where the shuttles live when they're not in space, where service is done, where the launchpads are. A landing at Edwards requires lashing the shuttle to a jetliner and flying it across the country to Kennedy afterward. The engineers had calculated the approach and concluded that Discovery had two shots at Kennedy, either on orbit number 135 or number 136 in order to make the scheduled return day. After that, it had two shots at Edwards. If either of those proved bad for some reason, Discovery could stay up another day and try again, but the weather for Sunday looked even worse in both places. Saturday, the scheduled day, clearly had to be it.

Bringing the space shuttle back to Earth is a complicated maneuver that, like liftoff, has certain points of no return. The shuttle doesn't fly in for a landing like some husky airliner, but instead falls out of orbit and glides in from hundreds of miles up. At thousands of miles an hour. There are no second chances at the runway, there is no circling around for a second pass, and there is no putting the shuttle back up into orbit once the decision has been made to break out and take a landing opportunity. So the question before NASA's engineers in Houston was essentially an unanswerable one: will the winds on the Florida runway get too strong for a safe landing after the shuttle has broken out of orbit and begun its seventy-minute descent to the ground?

Someone had to take a guess at that

From hero to legend, Glenn was in great demand upon his return.

TY GREENLEES

because the moment for making the landing on orbit 135 was drawing near. Discovery would need advance warning so Commander Brown could turn the shuttle around and get it facing backwards in its orbital path, the necessary position for firing the twin engines of the orbital maneuvering system. Burning those jets for several minutes would slow the shuttle down enough to let gravity take over and begin the long, fast pull to Earth. Just like the retrorockets slung beneath the belly of Glenn's old Mercury capsule. In shuttle language this is called the deorbit burn, and it would have to be done in the next few minutes if Discovery was going to make the first window to land at Kennedy. But the wind was on the edge of being too strong. The deorbit burn had to begin in nine minutes, so NASA had to decide right now. Either give Brown the go-ahead, or let orbit 135 go by, hope the wind slackens, and try again on orbit 136.

NASA's launch entry flight director in Houston waited and waited and waited for the wind to do something. Pick up steam or slow down, but do something that would make the decision easy. The wind refused to help, holding steady at just under seventeen m.p.h., just this side of being too strong. Finally, the entry director polled her mission control team and they all agreed: fire the jets for the deorbit burn and bring Discovery down to Kennedy.

It was as down-to-the-wire as NASA likes to go.

With that decision, the questions about John Glenn's fate upon returning to Earth slipped out of the academic like the shuttle coming out of orbit and became a concrete reality steaming inevitably toward an answer. Would John Glenn walk out of the shuttle? The answer was headed for Earth at Mach 24. Some time after the deorbit burn, gravity began, for the first time in nine days, pulling on Glenn's seventy-seven-year-old body. Fiery explosions encased the shuttle as it slammed back into the air molecules of Earth's atmosphere, but Glenn couldn't see any of it from his seat in the metal box called the mid-deck. He only could hear the popping, like old-fashioned flashbulbs. The ride is smooth. But gravity doesn't just ease back to normal force; the process of coming back to Earth brings a G-load that's one and a half times as strong as normal gravity. The blood tends to pool in your legs during reentry, which is why everyone stays seated for a few minutes after the shuttle stops on the runway. And making the adjustment to gravity is made even more difficult because the astronauts are wearing the eighty extra pounds built into their orange space suits.

"I've always kind of felt like a 100-year-old man," Parazynski said, "[in] the first moments getting back home."

Three sections of bleachers were filled out by the landing strip. Astronaut guests and friends mingled with NASA VIPs, including Administrator Dan Goldin who shielded his eyes from the sun with a baseball cap bearing the Discovery mission patch. Astronaut families had their own bleacher section further down the runway, closer to the end where the touchdown occurs. A nice crowd had gathered, and a fleet of buses rumbled their dirty diesel idle in a parking area not far away, but this was not the celebrity-infested gathering that had come to watch Glenn go up. Meanwhile, at this same time in the shuttle, Commander Brown was going through what he himself described as the "Oh my God!" part of the flight for every commander: the last 50,000 feet of altitude when computers aren't flying the shuttle, you are, and you realize this is not a simulation and you actually must land the shuttle for your six crewmates.

At high noon, the shuttle's tell-tale twin sonic booms burst from above, and the

TY GREENLEES

Glenn gives a "thumb up" during the welcome home celebration in Houston.

crowd, which had been craning their necks and straining their eyes to see it come in, cheered. Suddenly, with absolutely no warning, Discovery was big as a truck in the sky, coming straight toward the crowd, down for the pavement. At 12:03 p.m. on Saturday, November 7, 1998, the rear wheels of Discovery barked against the hot pavement at the Kennedy Space Center's Runway 33. Thirty seconds later, the nose came down, and by 12:04:29 p.m. Discovery was at rest. The crosswinds had held steady, just inside the safety threshold.

But the day was not over for the crew. Glenn had worn an elaborate heart monitor under his launch and reentry suit so he could be part of an ongoing study of astronauts whose cardiovascular systems are tracked during the descent from orbit. After being unhooked from that, Glenn faced more than four hours of medical examination so doctors could chart his body's every condition as it adjusted to the shock of being reintroduced to gravity in one grand swoop. But first, the finite question of Glenn's future would have to be addressed. The walk. Could he? Two full hours passed while the crew moved from the shuttle into a transport vehicle, which is a bus-shaped truck with enormous hydraulic legs that lift the entire crew compartment up to the level of the shuttle's hatch. The astronauts tugged themselves free of their launch suits, spruced up a bit, and prepared for the world.

John Glenn wasn't quite ready. The answer to the question of whether he would walk off the shuttle was undoubtedly, yes. Yes, he walked off the shuttle. But it also was clear from his gentle, squared gait that he still was suffering from an astronaut version of the bed spins. His inner ear, which controls balance, had adapted well to the weightless world where there is no up or down, but now it seemed a little reluctant to switch back. The NASA tradition of walking around the shuttle on the runway was shortened, but they did it, and Glenn would only need some time to adjust. A good shower — not too hot, he was warned, they've had astronauts faint in steamy showers — some lasagna and then a little Doublemint chewing gum really helped. A crew press conference for late that evening was canceled so the astronauts wouldn't have too long a day. (It had started at 4:30 a.m., and the press conference was looking more and more like an 8:30 p.m. event as doctors finished gathering their data.)

One night made all the difference. Glenn arrived at the morning press conference with his crewmates fresh and energetic, sure-footed and extremely happy. He quipped and joked along with the rest of the crew through a thirty-minute Q&A in which he didn't dodge the obvious question about how well he had done the day before.

"I didn't feel too hot when I got off [the shuttle], Glenn said. "Obviously I was walking a little spraddle-legged yesterday to keep my balance a little bit, but everything went very well."

Asked again, he answered again. "I think when you come back from something like this you feel maybe a little bit woozy when you come back in. I'm probably ninety-five or ninety-eight percent back to normal now. I'm not just whipping my head around quite like that yet, but that'll come over a day or so, I guess. And, I feel great."

He admitted the shuttle inspection or "walk around," as they call it, wasn't much fun. But no, he said, he felt no pressure at all to participate in the walk because the eyes of the world were upon him. He had not insisted he walk around the shuttle on behalf of senior citizens or space enthusiasts or anyone else who had loved the symbolism of his

return to space. Glenn's determination to wobble about with the other astronauts, much less the worse for wear than he, was driven by the simple truth that that's what astronaut crews do.

"I wanted to get out with the rest of the crew and do the walk around, and . . . if I'd have been on my hands and knees I was going to do it," he said before adding with a smile, "I wasn't quite to that point."

A camera on a tripod in the back of the room toppled, setting off dominoes of noisy photo gear, and Glenn yelled out, "This is the dangerous part!" which drew applause. When a network reporter began his question by reciting the extraordinary achievements in Glenn's life, and when it sounded like the compliments were winding down and the question was coming up, the usually modest Glenn chirped, "Keep talking".

But all in all, the fun part was over. There would be weeks yet of data collection. More needle sticks, a few more nights trying to sleep in that "contraption," as Annie called the electrode net used in the sleep study, and on and on. There might be some future relationship with NASA, a consultant job, maybe, or advisor on some issues. But there wouldn't be another space flight. For the second time in his life, John Glenn faced the notion that he would never fly again in space, although it is entirely possible this actually was the first time he saw it as a kind of immovable truth. Obviously, he hadn't given up on the possibility when he left the space program in 1964.

"We'd like to make STS-95 B and go right back up again, but that's not to be," he said, trying to assess the reality of a dream achieved. "A sense of accomplishment, I guess I feel. And a little bit let down that the whole thing is over maybe. But nothing serious, I'm quite elated how the whole thing went."

What would be next for him in the grand,

Glenn talks about his 1998 mission against the backdrop of his 1962 flight.

wide version of that question remained open. Ohio State University had created the John Glenn Institute for Public Policy and Public Service just before his shuttle flight, and he had agreed to send his Senate papers there and teach seminars on government. He also had been trying throughout the training and flying of the space shuttle mission to record his thoughts into a tape recorder, perhaps for a book, but he'd get involved in the mission itself and forget sometimes. He even went so far as to Velcro the recorder above his bunk in the shuttle to remind him.

The question of a third space flight came up.

"You'd have to check with Annie on that. She's rather firm in her attitude on that at the moment. But probably not. I would

imagine this will probably be my last flight," he said, before opening a grin and adding, "unless there's some rising demand that I go back up again."

From where? The only rising demand that ever mattered in John Glenn's life was the one inside of him. But it mattered two or three times more than it does in most of the rest of us. The right place at the right time was more than luck in John Glenn's life, and it was more than coincidence. NASA wouldn't have sent an elderly man into space if John Glenn hadn't pushed them into it, any more than the Marines would have set a transcontinental speed record in 1957 or the Mercury Program would have taken a redhead from Ohio who didn't have the engineering degree they expected of all their astronauts. The obvious lesson from John Glenn's return to space was that age need not be an obstacle to dreams, but it runs much deeper than that. His space shuttle journey was not some anomaly in a life of modest gains; it was just another day of extraordinary achievement in a lifetime filled with them.

Traveling 3.5 million miles in nine days aboard the space shuttle at the age of seventy-seven simply is what you do when you're John Glenn.